TWILIGHT OF THE MACHINES

JOHN ZERZAN

Feral House
1240 W. Sims Way #124
Port Townsend WA 98368

www.FeralHouse.com

Book Design by Lissi Erwin/Splendid Corp.

TWILIGHT OF
THE MACHINES

JOHN ZERZAN

TABLE OF CONTENTS

Preface

Specialization, domestication, civilization, mass society, modernity, technoculture... behold Progress, its fruition presented more and more unmistakably. The imperative of control unfolds starkly, pushing us to ask questions equal to the mounting threat around us and within us. These dire times may yet reveal invigorating new vistas of thought and action. When everything is at stake, all must be confronted and superseded. At this moment, there is the distinct possibility of doing just that.

People all over the world are showing that they are ready to engage in this dialogue. The challenge is to broach a new conversation in one's own society. The effort begins with a refusal to accept the givens that are turning on us so relentlessly, so viciously. The confrontation is with the increasingly pathological state of modern society: outbursts of mass homicide, an ever more drug-reliant populace, amid a collapsing physical environment. The initiative remains disconnected and marginalized, with enormous inertia and denial in its way. But reality is persistent, and it's calling forth a questioning that is as unprecedented as the darkening situation we face.

Clinging to politics is one way of avoiding the confrontation with the devouring logic of civilization, holding instead with the accepted assumptions and definitions. Leaving it all behind is the opposite: a truly qualitative change, a fundamental paradigm shift.

This change is *not* about:

- seeking "alternative" energy sources to power all the projects and systems that should never have been started up in the first place;
- being vaguely "post-Left", the disguise that some adopt while changing none of their (leftist) orientations;
- espousing an "anti-globalization" orientation that's anything but, given activists' near-universal embrace of the totalizing industrial world system;
- preserving the technological order, while ignoring the degradation of millions and the systematic destruction of the earth that undergird the existence of every part of the technoculture;
- claiming—as anarchists—to oppose the state, while ignoring the fact that this hypercomplex global setup couldn't function for a day without many levels of government.

The way is open for radical change. If complex society is itself the issue, if class society began with division of labor in the Neolithic, and if the Brave New World now moving forward was born with the shift to domesticated life, then all we've taken for granted is implicated. We are seeing more deeply, and the explorations must extend to include everyone. A daunting, but exciting opportunity!

Twilight of the Machines is offered in this spirit. Part 1 deals with remote origins and developments within early civilization. Part 2 has a more contemporary focus. May assumptions be questioned, and may conversations proliferate!

—John Zerzan

Part I:
Origins of
the Crisis

Too Marvelous for Words
(LANGUAGE BRIEFLY REVISITED)

A few years ago the now-deceased philosopher of science and anarchist Paul Feyerabend was invited to sign a petition being circulated by well-known European thinkers. Its thrust was that society is in need of input from philosophers, who draw upon the "intellectual treasures" of the past. In these dark times, the petition concluded, "We need philosophy."

Derrida, Ricoeur and the other liberal concocters of the document were no doubt shocked by Feyerabend's negative reaction. He pointed out that philosophy's "treasures" were not meant as additions to ways of living, but were intended to express their replacement. "Philosophers," he explained, "have destroyed what they have found, much in the way that the [other] standard-bearers of Western civilization have destroyed indigenous cultures. . ."[1] Feyerabend wondered how civilized rationality—which has reduced a natural abundance of life and freedom and thereby devalued human existence—became so dominant. Perhaps its chief weapon is symbolic thought, with its ascendancy in the form of language. Maybe the wrong turn we took as a species can be located at that milestone in our evolution.

"Writing. . . can be seen to *cause a new reality to come into being*," according to Terence Hawkes, who adds that language "allows no single, unitary appeals to a 'reality' beyond itself. In the end, it constitutes its own reality."[2] An infinitely diverse reality is captured by finite language; it subordinates all of nature to its formal system. As Michael Baxandall put it, "Any language. . . is a conspiracy against experience in the sense of being a collective attempt to simplify and arrange experience into manageable parcels."³

At the beginning of domination and repression, the start of the long process of depleting the riches of the living world, is a very ill-advised separation from the flow of life. What was once freely given is now controlled, rationed, distributed. Feyerabend refers to the effort, especially by specialists, to "reduce the abundance that surrounds and confuses them."⁴

The essence of language is the symbol. Always a substitution. Always a paler re-presentation of what is at hand, what presents itself directly to us. Susanne Langer pondered the mysterious nature of symbols: "If the word 'plenty' were replaced by a succulent, real, ripe peach, few people could attend to the mere content of the word. The more barren and indifferent the symbol, the greater its semantic power. Peaches are too good to act as words; we're too much interested in peaches themselves."⁵

For the Murngin people of northern Australia, name giving and all other such linguistic externalizations are treated as a kind of death, the loss of an original wholeness. This is very much to the point of what language itself accomplishes. In slightly more general terms, Ernest Jones proposed that "only what is repressed is symbolized; only what is repressed needs to be symbolized."⁶

Any symbolic mode is only one way of seeing and connecting. By reversing our steps, in light of what has been progressively de-realized or lost, it appears likely that before the symbolic dimension took over, relations between people were more subtle, unmediated, and sensual. But this is a forbidden notion. Commonplace statements like: "Verbal language was perhaps the greatest technical invention [!] of human life" and "Language enables human beings to communicate and share with each other" deny, incredibly, that communication, sharing, society didn't exist before the symbolic, which was such a relative late-comer on the evolutionary scale. (It appeared an estimated 35,000 years ago, following nearly two million years of successful human adaptations to life on earth.) Such formulations express perfectly the hubris, imperialism and ignorance of symbolic thought.

We don't know when speech originated; but soon after domestication gained the upper hand over foraging or gatherer-hunter life, writing appeared. By about 4500 B.C. engraved clay tokens, records of agricultural transactions and inventories, became widespread in the Middle East. Five thousand years later, the Greek perfection of the alphabet completed the transition to modern writing systems.

The singular excellence of modern humanshas of course become a basic tenet of civilization's ideology. It extends, for example, to Sapir's definition of personality as a systematic psychological organization depending on constellations of symbols.[7] The symbolic medium of language is now widely felt as an all-defining imprisonment, rather than a liberatory triumph. A great deal of philosophical analysis in the past century revolves around this realization, though we can hardly imagine breaking free of it or even clearly recognizing its pervasive presence and influence. This is a measure of the depth of the impoverishing logic that Feyerabend sought to understand.Certainly it is no small endeavor to try to imagine what human cognition may have been like, before language and symbolic thought took possession of so much of our consciousness.

It is grammar that establishes language as a system, reminding us that the symbolic must become systemic in order to seize and hold power. This is how the perceived world becomes structured, its abundance processed and reduced. The grammar of every language is a theory of experience, and more than that, it's an ideology. It sets rules and limits, and grinds the one-prescription-fits-all lenses through which we see everything. A language is defined by grammatical rules (not of the speaker's choosing); the human mind is now commonly seen as a grammar- or syntax-driven machine. As early as the 1700s, human nature was described as "a tissue of language,"[8] a further measure of the hegemony of language as the determining ground of consciousness.

Language, and symbolism in general, are always substitutive, implying meanings that cannot be derived directly from experiential contexts. Here is the long-ago source of today's generalized crisis of meaning. Language initiates and reproduces a distinction or separation that leads to ever-increasing place-lessness. Resistance to this impoverishing movement must lead to the problematization of language. Foucault noted that speech is not merely "a verbalization of conflicts and systems of domination, but...the very object of man's conflicts." He didn't develop this point, which is valid and deserves our attention and study. The roots of today's globalizing spiritual crisis lie in a movement away from immediacy; this is the hallmark of the symbolic.

Civilization has made repeated, futile efforts to overcome the instability and erosion of substance caused by the rule of the symbolic. Among the most well-known was Descartes' attempt to give "grounding" to science and modernity in the 17th century. His famous mind-body duality provides a philosophical method (based on suppression of the body, of course) that we have suffered from ever since. He claimed certainty for the system by means of the language of number, as expressed in his analytic geometry. But the dream of certainty has been consistently revealed as a further repressive substitute: an illusory foundation on which domination has extended itself in every direction.

Language is conformist in the profoundest sense; even objective reality yields to its pressure. The so-called factual is brought to dissolution, because it is shaped and constrained by the limits of language. Under its reductive force, we forget that we don't need symbols to be present to meaning. The reality of pre-linguistic social practices is screened from us by more than the practical, empirical limitations of access to time past. Primal existence has been ruled irrelevant, and indigenous lifeways are everywhere under siege, because of civilization's pervasive over-valuation of the symbolic.

Yet an exploration of social life in the early symbolic epoch need not be overly speculative, and may reveal important connections. We know from archaeological and ethnographic evidence that early on in divided society, inequality was often based on ritual knowledge: who possessed it, who did not. The symbolic must have already been very much present and determinant; or why wouldn't inequality be based on, say, knowledge of plants?

It could well be that language emerged from ritual, which among other attributes, is a substitutive form of emotion. The dissociated, symbolic process of ritual activity parallels that of language and may have first generated it: emotionally displaced expression, abstracted cries; language as ritualized expression.

From early on, ritual has mystified power relationships. Deacon has argued that language became necessary to enable the contracts on which society depends.[10] However, it is more than likely that social life long predated language. Contracts based on language may have appeared to meet some challenge in society, such as the beginnings of disequilibrium or inequality.

At a later stage, religion was a further (and even less successful) response to problems and tensions in human communities. Language was central there, too. Word magic runs through the history of religions; veneration of names and naming is common (the history of religious life in Ancient Egypt is a well-documented example).[11]

Problems introduced by complexity or hierarchy have never been resolved by symbolic means. What is overcome symbolically remains intact on the non-symbolic (real) plane. Symbolic means sidestep reality; they are part of what is going wrong. Division of labor, for instance, eroded face-to-face interaction and eroded people's direct, intimate relationship with the natural world. The symbolic is complicit; it generates more and more mediations to accompany those created by social practices. Life becomes fragmented; connections to nature are obscured and dissolved. Instead of repairing the rupture, symbolic thought turns people in the wrong direction: toward abstraction. The "thirst for transcendence" is initiated, ignoring the shifting reality that created that desire in the first place. Language plays a key role here, re-ordering and subordinating natural systems that humankind was once attuned to. Symbolic culture demands that we reject our "animal nature" in favor of a symbolically defined "human nature".

Now we live our everyday lives in a world system that is ever more symbolic and disembodied. Even economies are decisively symbolic; and we are told that the social bond (what's left of it) is essentially linguistic. Language was an intrusion that brought on a series of transformations resulting in our loss of the world. Once, as Freud put it, "the whole world was animate,"[12] known by all in a full, engaged way. Later the totem animal was replaced by a god, a signpost of the advancing symbolic. (I am reminded that indigenous elders who are asked to make audio or video recordings often decline, insisting that what they say must be communicated in person, face to face.)

Language was a powerful instrument for technological and social disenchantment. Like every symbolic device, it was itself an invention. But it does not establish or generate meaning, which antedates language. Rather, it confines and distorts meaning, via the rules of symbolic representation—the architecture of the logic of control. Domestication also partakes of this underlying orientation, which has served domination in key ways. Language has a standardizing quality; this develops in tandem with the technological development it facilitates. The printing press, for example, suppressed dialects and other language variants, creating unified standards for exchange and communication. Literacy has always served economic development, and aimed to bolster the cohesion so necessary for the nation-state and nationalism.

Language is a productive force; like technology, it is not amenable to social control. In the postmodern era, both language and technology rule, but each shows signs of exhaustion. Today's symbolic reflects nothing much more than the habit of power behind it. Human connectedness and corporeal immediacy have been traded away for a fading sense of reality. The poverty and manipulation of mass communication is the postmodern version of culture. Here is the voice of industrial modernity as it goes cyber/digital/ virtual, mirroring its domesticated core, a facet of mass production.

Language does not bestow presence; rather, it banishes presence and its transparency. We are "condemned to words," said Marlene Nourbese Philip. She provides a wonderful metaphor of origins:

> God first created silence: whole, indivisible, complete. All
> creatures—man, woman, beast, insect, bird and fish—lived

happily together with this silence, until one day man and
woman lay down together and between them created the first
word. This displeased God deeply and in anger she shook
out her bag of words over the world, sprinkling and shower-
ing her creation with them. Her word store rained down
upon all creatures, shattering forever the whole that once was
silence. God cursed the world with words and forever after
it would be a struggle for man and woman to return to the
original silence.[13]

Dan Sperber wrote of an "epidemiology of representations"; his
pathology metaphor is apt. He questioned why the symbolic spreads
like an epidemic, why we are susceptible to it,[14] but left these questions
unanswered.[15]

In the Age of Communication our homogenized symbolic "materi-
als" prove so inadequate. Our isolation grows; what we have to communicate shrinks. How is it that the world and consciousness have
come to be seem as mainly comprised of, and enclosed by, language?
Does time structure language or does language structure time? So many
questions, including the key one; how do we transcend, escape, get rid
of the symbolic?

We may not yet know much about the how, but at least we know
something of the why. In language, number, art, and the rest, a substitution essence has been the symbolic's bad bargain. This compensa-
tion fails to compensate for what is surrendered. Symbolic transactions
deliver an arid, anti-spiritual dimension, emptier and colder with each
re-enactment. This is nothing new; it's just more sadly oppressive and
obvious, more corrosive of actual connectedness, particularity, non-
programmed life. This strangling, unhappy state saps our vitality and
will destroy us if we don't end it.

Representation is unfaithful even to itself. Geert Lovink concluded
that "there is no 'natural' image anymore. All information has gone
through the process of digitization. We just have to deal with the fact
that we can no longer believe our eyes, our ears. Everyone who has

worked with a computer will know this."[16] Discounted, atrophying senses to go along with the distancing and decontextualization.

George Steiner has announced a "core tiredness" as the climate of spirit today. The weight of language and the symbolic has brought this fatigue; the "shadows lengthen" and there is "valediction in the air."[17] A farewell is indeed appropriate. Growing illiteracy, cheapened channels of the symbolic (e.g. email)...a tattered dimension. The Tower of Babel, now built into cyberspace, has never been taller—but quite possibly never so weakly supported. Easier to bring down?

2

Patriarchy, Civilization and the Origins of Gender

Civilization, very fundamentally, is the history of the domination of nature and of women. Patriarchy means rule over women and nature. Are the two institutions at base synonymous?

Philosophy has mainly ignored the vast realm of suffering that has unfolded since it began, in division of labor, its long course. Hélène Cixous calls the history of philosophy a "chain of fathers." Women are as absent from it as suffering, and are certainly the closest of kin.

Camille Paglia, anti-feminist literary theorist, meditates thusly on civilization and women:

> When I see a giant crane passing on a flatbed truck, I pause
> in awe and reverence, as one would for a church procession.
> What power of conception: what grandiosity: these cranes
> tie us to ancient Egypt, where monumental architecture was
> first imagined and achieved. If civilization has been left in
> female hands, we would still be living in grass huts. [1]

The "glories" of civilization and women's disinterest in them. To some of us the "grass huts" represent not taking the wrong path, that of oppression and destructiveness. In light of the globally metastasizing death drive of technological civilization, if only we still lived in grass huts!

Women and nature are universally devalued by the dominant paradigm and who cannot see what this has wrought? Ursula Le Guin gives us a healthy corrective to Paglia's dismissal of both:

> Civilized Man says: I am Self, I am Master, all the rest is
> other—outside, below, underneath, subservient. I own, I
> use, I explore, I exploit, I control. What I do is what matters.
> What I want is what matter is for. I am that I am, and the rest
> is women and wilderness, to be used as I see fit. [2]

There are certainly many who believe that early civilizations existed that were matriarchal. But no anthropologists or archaeologists, feminists included, have found evidence of such societies. "The search for a genuinely egalitarian, let along matriarchal, culture has proved fruitless," concludes Sherry Ortner.[3]

There was, however, a long span of time when women were generally less subject to men, before male-defined culture became fixed or universal. Since the 1970s anthropologists such as Adrienne Zihlman, Nancy Tanner and Frances Dahlberg[4] have corrected the earlier focus or stereotype of prehistoric "Man the Hunter" to that of "Woman the Gatherer." Key here is the datum that as a general average, pre-agricultural band societies received about 80 percent of their sustenance from gathering and 20 percent from hunting. It is possible to overstate the hunting/gathering distinction and to over-look those groups in which, to significant degrees, women have hunted and men have gathered.[5] But women's autonomy in foraging societies is rooted in the fact that material resources for subsistence are equally available to women and men in their respective spheres of activity.

In the context of the generally egalitarian ethos of hunter-gatherer or foraging societies, anthropologists like Eleanor Leacock, Patricia Draper and Mina Caulfield have described a generally equal relationship between men and women.[6] In such settings where the person who procures something also distributes it and where women procure about 80 percent of the sustenance, it is largely women who determine band society movements and camp locations. Similarly, evidence indicates that both women and men made the stone tools used by pre-agricultural peoples. [7]

With the matrilocal Pueblo, Iroquois, Crow, and other American Indian groups, women could terminate a marital relationship at any time. Overall, males and females in band society move freely and peacefully from one band to another as well as into or out of relationships.[8] According to Rosalind Miles, the men not only do not command or exploit women's labor, "they exert little or no control over women's bodies or those of their children, making no fetish of virginity or chastity, and making no demands of women's sexual exclusivity."[9] Zubeeda Banu Quraishy provides an African example: "Mbuti gender associations were characterized by harmony and cooperation."[10]

And yet, one wonders, was the situation really ever quite this rosy? Given an apparently universal devaluation of women, which varies in its forms but not in its essence, the question of when and how it was basically otherwise persists. There is a fundamental division of social existence according to gender, and an obvious hierarchy to this divide. For philosopher Jane Flax, the most deep-seated dualisms, even including those of subject-object and mind-body, are a reflection of gender disunity.[11]

Gender is not the same as the natural/physiological distinction between the sexes. It is a cultural categorization and ranking grounded in a sexual division of labor that may be the single cultural form of greatest significance. If gender introduces and legitimates inequality and domination, what could be more important to put into question? So in terms of origins—and in terms of our future—the question of human society without gender presents itself.

We know that division of labor led to domestication and civilization, and drives the globalized system of domination today. It also appears that artificially imposed sexual division of labor was its earliest form and was also, in effect, the formation of gender.

Sharing food has long been recognized as a hallmark of the foraging life-way. Sharing the responsibility for the care of offspring, too, which can still be seen among the few remaining hunter-gatherer societies, in contrast to privatized, isolated family life in civilization. What we think of as the family is not an eternal institution, any more than exclusively female mothering was inevitable in human evolution.[12]

Society is integrated via the division of labor and the family is integrated via the sexual division of labor. The need for integration bespeaks a tension, a split that calls for a basis for cohesion or solidarity. In this sense Testart is right: "Inherent in kinship is hierarchy."[13] And with their basis in division of labor, the relations of kinship become relations of production. "Gender is inherent in the very nature of kinship," as Cucchiari points out, "which could not exist without it." It is in this area that the root of the domination of nature as well as of women may be explored.

As combined group foraging in band societies gave way to specialized roles, kinship structures formed the infrastructure of relationships that developed in the direction of inequality and power differentials. Women typically became immobilized by a privatizing child care role; this pattern deepened later on, beyond the supposed requirements of that gender role. This gender-based separation and division of labor began to occur around the transition from the Middle to Upper Paleolithic eras.[15]

Gender and the kinship system are cultural constructs set over and against the biological subjects involved, "above all a symbolic organization of behavior," according to Juliet Mitchell.[16] It may be more telling to look at symbolic culture itself as required by gendered society, by "the need to mediate symbolically a severely dichotomized cosmos."[17] The which-came-first question introduces itself, and is difficult to resolve. It is clear, however, that there is no evidence of symbolic activity (e.g. cave paintings) until the gender system, based on sexual division of labor, was apparently under way.[18]

By the Upper Paleolithic, that epoch immediately prior to the Neolithic Revolution of domestication and civilization, the gender revolution had won the day. Masculine and feminine signs are present in the first cave art, about 35,000 years ago. Gender consciousness arises as an all-encompassing ensemble of dualities, a specter of divided society. In the new polarization activity becomes gender-related, gender-defined. The role of hunter, for example, develops into association with males, its requirements attributed to the male gender as desired traits.

That which had been far more unitary or generalized, such as group foraging or communal responsibility for child tending, had now become the separated spheres in which sexual jealousy and possessiveness appear. At the same time, the symbolic emerges as a separate sphere or reality. This is revealing in terms of the content of art, as well as ritual and its practice. It is hazardous to extrapolate from the present to the remote past, yet surviving non-industrialized cultures may shed some light. The Bimin-Kushusmin of Papua New Guinea, for example, experience the masculine-feminine split as fundamental and defining. The masculine "essence," called *finiik*, not only signifies powerful, warlike qualities but also those of ritual and control. The feminine "essence," or *khaapkhabuurien*, is wild, impulsive, sensuous, and ignorant of ritual.[19] Similarly, the Mansi of northwestern Siberia place severe restrictions on women's involvement in their ritual practices.[20] With band societies, it is no exaggeration to say that the presence or absence of ritual is crucial to the question of the subordination of women. Gayle Rubin concludes that the "world-historical defeat of women occurred with the origins of culture and is a prerequisite of culture."[22]

The simultaneous rise of symbolic culture and gendered life is not a coincidence. Each of them involves a basic shift from non-separated, non-hierarchized life. The logic of their development and extension is a response to tensions and inequalities that they incarnate; both are dialectically interconnected to earliest, artificial division of labor.

On the heels, relatively speaking, of the gender/symbolic alteration came another Great Leap Forward, into agriculture and civilization. This is the definitive "rising above nature," overriding the previous two million years of non-dominating intelligence and intimacy with nature. This change was decisive as a consolidation and intensification of the division of labor. Meillasoux reminds us of its beginnings:

> Nothing in nature explains the sexual division of labor, nor
> such institutions as marriage, conjugality or paternal filia-
> tion. All are imposed on women by constraint, all are there-
> fore facts of civilization which must be explained, not used
> as explanations. [23]

Kelkar and Nathan, for example, did not find very much gender specialization among hunter-gatherers in western India, compared to agriculturalists there. [24] The transition from foraging to food production brought similar radical changes in societies everywhere. It is instructive, to cite another example closer to the present, that the Muskogee people of the American Southeast upheld the intrinsic value of the untamed, undomesticated forest; colonial civilizers attacked this stance by trying to replace Muskogee matrilineal tradition with patrilineal relations. [25]

The locus of the transformation of the wild to the cultural is the domicile, as women become progressively limited to its horizons. Domestication is grounded here (etymologically as well, from the Latin *domus*, or household): drudge work, less robusticity than with foraging, many more children, and a lower life expectancy than males are among the features of agricultural existence for women. [26] Here another dichotomy appears, the distinction between work and non-work, which for so many, many generations did not exist. From the gendered production site and its constant extension come further foundations of our culture and mentality.

Confined, if not fully pacified, women are defined as passive. Like nature, of value as something to be made to produce; awaiting fertil-ization, activation from outside herself/ itself. Women experience the move from autonomy and relative equality in small, mobile anarchic groups to controlled status in large, complex governed settlements.

Mythology and religion, compensations of divided society, testify to the reduced position of women. In Homer's Greece, fallow land (not domesticated by grain culture) was considered feminine, the abode of Calypso, of Circe, of the Sirens who tempted Odysseus to abandon civilization's labors. Both land and women are again subjects of domination. But this imperialism betrays traces of guilty conscience, as in the punishments for those associated with domestication and technology, in the tales of Prometheus and Sisyphus. The project of agriculture was felt, in some areas more than others, as a violation; hence, the incidence of rape in the stories of Demeter. Over time as the losses mount, the great mother-daughter relationships of Greek myth—Demeter-Kore, Clytemnestra-Iphigenia, Jocasta-Antigone, for example—disappear.

In Genesis, the Bible's first book, woman is born from the body of man. The Fall from Eden represents the demise of hunter-gatherer life, the expulsion into agriculture and hard labor. It is blamed on Eve, of course, who bears the stigma of the Fall. [27] Quite an irony, in that domestication is the fear and refusal of nature and woman, while the Garden myth blames the chief victim of its scenario, in reality.

Agriculture is a conquest that fulfills what began with gender formation and development. Despite the presence of goddess figures, wedded to the touchstone of fertility, in general Neolithic culture is very concerned with virility. From the emotional dimensions of this masculinism, as Cauvin sees it, animal domestication must have been principally a male initiative. [28] The distancing and power emphasis have been with us ever since; frontier expansion, for instance, as male energy subduing female nature, one frontier after another.

This trajectory has reached overwhelming proportions, and we are told on all sides that we cannot avoid our engagement with ubiquitous technology. But patriarchy too is everywhere, and once again the inferiority of nature is presumed. Fortunately "many feminists," says Carol Stabile, hold that "a rejection of technology is fundamentally identical to a rejection of patriarchy." [29]

There are other feminists who claim a part of the technological enterprise, which posits a virtual, cyborg "escape from the body" and its gendered history of subjugation. But this flight is illusory, a forgetting of the whole train and logic of oppressive institutions that make up patriarchy. The dis-embodied high-tech future can only be more of the same destructive course.

Freud considered taking one's place as a gendered subject to be foundational, both culturally and psychologically. But his theories assume an already present gendered subjectivity, and thus beg many questions. Various considerations remain unaddressed, such as gender as an expression of power relations, and the fact that we enter this world as bisexual creatures.

Carla Freeman poses a pertinent question with her essay titled, "Is Local: Global as Feminine: Masculine? Rethinking the Gender of Globalization".[30]

The general crisis of modernity has its roots in the imposition of gender. Separation and inequality begin here at the period when symbolic culture itself emerges, soon becoming definitive as domestication and civilization: patriarchy. The hierarchy of gender can no more be reformed than the class system or globalization. Without a deeply radical women's liberation we are consigned to the deadly swindle and mutilation now dealing out a fearful toll everywhere. The wholeness of original genderlessness may be a prescription for our redemption.

3

On the Origins of War

War is a staple of civilization. Its mass, rationalized, chronic presence has increased as civilization has spread and deepened. Among the specific reasons it doesn't go away is the desire to escape the horror of mass-industrial life. Mass society of course finds its reflection in mass soldiery and it has been this way from early civilization. In the age of hyper-developing technology, war is fed by new heights of dissociation and disembodiment. We are ever further from a grounding or leverage from which to oppose it (while too many accept paltry, symbolic "protest" gestures).

How did it come to be that war is "the proper work of man," in the words of Homer's Odysseus? We know that organized warfare advanced with early industry and complex social organization in general, but the question of origins predates even Homer's early Iron Age. The explicit archaeological/anthropological literature on the subject is surprisingly slight.

Civilization has always had a basic interest in holding its subjects captive by touting the necessity of official armed force. It is a prime ideological claim that without the state's monopoly on violence, we would be unprotected and insecure. After all, according to Hobbes, the human condition has been and will always be that of "a war of all against all." Modern voices, too, have argued that humans are innately aggressive and violent, and so need to be constrained by armed authority. Raymond Dart (e.g. *Adventures with the Missing Link*, 1959), Robert Ardrey (e.g. *African Genesis*, 1961), and Konrad Lorenz (e.g. *On Aggression*, 1966) are among the best known, but the evidence they put forth has been very largely discredited.

In the second half of the 20th century, this pessimistic view of human nature began to shift. Based on archaeological evidence, it is now a tenet of mainstream scholarship that pre-civilization humans lived in the absence of violence—more specifically, of organized violence. Eibl-Eibesfeldt referred to the !Ko-Bushmen as not bellicose: "Their cultural ideal is peaceful coexistence, and they achieve this by avoiding conflict, that is by splitting up, and by emphasizing and encouraging the numerous patterns of bonding."[1] An earlier judgment by W.J. Perry is generally accurate, if somewhat idealized: "Warfare, immorality, vice, polygyny, slavery, and the subjection of women seem to be absent among our gatherer-hunter ancestors."[2]

The current literature consistently reports that until the final stages of the Paleolithic Age—until just prior to the present 10,000-year era of domestication—there is no conclusive evidence that any tools or hunting weapons were used against humans at all.[3] "Depictions of battle scenes, skirmishes and hand-to-hand combat are rare in hunter-gatherer art and when they do occur most often result from contact with agriculturalists or industrialized invaders," concludes Taçon and Chippindale's study of Australian rock art.[4] When conflict began to emerge, encounters rarely lasted more than half an hour, and if a death occurred both parties would retire at once.[5]

The record of Native Americans in California is similar. Kroeber reported that their fighting was "notably bloodless. They even went so far as to take poorer arrows to war than they used in economic hunting."[6] Wintu people of Northern California called off hostilities once someone was injured. "Most Californians were absolutely nonmilitary; they possessed next to none of the traits requisite for the military horizon, a condition that would have taxed their all but nonexistent social organization too much. Their societies made no provision for collective political action," in the view of Turney-High.[8] Lorna Marshall described Kung! Bushmen as celebrating no valiant heroes or tales of battle. One of them remarked, "Fighting is very dangerous; someone might get killed."[9] George Bird Grinnell's "Coup and Scalp Among the Plains Indians"[10] argues that counting coup (striking or touching an enemy with the hand or a small stick) was the highest point of (essentially nonviolent) bravery, whereas scalping was not valued.

The emergence of institutionalized warfare appears to be associated with domestication, and/or a drastic change in a society's physical situation. As Glassman puts it, this comes about "only where band peoples have been drawn into the warfare of horticulturalists or herders, or driven into an ever-diminishing territory."[11] The first reliable archaeological evidence of warfare is that of fortified, pre-Biblical Jericho, c. 7500 B.C. In the early Neolithic a relatively sudden shift happened. What dynamic forces may have led people to adopt war as a social institution? To date, this question has not been explored in any depth.

Symbolic culture appears to have emerged in the Upper Paleolithic; by the Neolithic it was firmly established in human cultures everywhere. The symbolic has a way of effacing particularity, reducing human presence in its specific, non-mediated aspects. It is easier to direct violence against a faceless enemy who represents some officially defined evil or threat. Ritual is the earliest known form of purposive symbolic activity: symbolism acting in the world. Archaeological evidence suggests that there may be a link between ritual and the emergence of organized warfare.

During the almost timeless era when humans were not interested in dominating their surroundings, certain places were special and came to be known as sacred sites. This was based on a spiritual and emotional kinship with the land, expressed in various forms of totemism or custodianship. Ritual begins to appear, but is not central to band or forager societies. Emma Blake observes, "Although the peoples of the Paleolithic practiced rituals, the richest material residues date from the Neolithic period onward, when sedentism and the domestication of plants and animals brought changes to the outlook and cosmology of people everywhere."[12] It was in the Upper Paleolithic that certain strains and tensions caused by the development of specialization first became evident. Inequities can be measured by such evidence as differing amounts of goods at hearth sites in encampments; in response, ritual appears to have begun to play a greater social role. As many have noted, ritual in this context is a way of addressing deficiencies of cohesion or solidarity; it is a means of guaranteeing a social order that has become problematic. As Bruce Knauft saw, "ritual reinforces and puts beyond argument or question certain highly general propositions about the spiritual and human world...[and] predisposes deep-seated cognitive acceptance and behavioral compliance with these cosmological propositions."[13] Ritual thus provides the original ideological glue for societies now in need of such legitimating assistance. Face-to-face solutions become ineffective as social solutions, when communities become complex and already partly stratified. The symbolic is a non-solution; in fact, it is a type of enforcer of relationships and world-views characterized by inequality and estrangement.

Ritual is itself a type of power, an early, pre-state form of politics. Among the Maring people of Papua New Guinea, for instance, the conventions of the ritual cycle specify duties or roles in the absence of explicitly political authorities. Sanctity is therefore a functional alternative to politics; sacred conventions, in effect, govern society.[14] Ritualization is clearly an early strategic arena for the incorporation of power relations. Further, warfare can be a sacred undertaking, with militarism promoted ritually, blessing emergent social hierarchy.

René Girard proposes that rituals of sacrifice are a necessary counter to endemic aggression and violence in society.[15] Something nearer to the reverse is more the case: ritual legitimates and enacts violence. As Lienhardt said of the Dinka herders of Africa, to "make a feast or sacrifice often implies war." Ritual does not substitute for war, according to Arkush and Stanish: "warfare in all times and places has ritual elements."[17] They see the dichotomy between "ritual battle" and "real war" to be false, summarizing that "archaeologists can expect destructive warfare and ritual to go hand in hand."[18]

It is not only that among Apache groups, for example, that the most ritualized were the most agricultural,[19] but that so often ritual has mainly to do with agriculture and warfare, which are often very closely linked.[20] It is not uncommon to find warfare itself seen as a means of enhancing the fertility of cultivated ground. Ritual regulation of production and belligerence means that domestication has become the decisive factor. "The emergence of systematic warfare, fortifications, and weapons of destruction," says Hassan, "follows the path of agriculture."[21]

Ritual evolves into religious systems, the gods come forth, sacrifice is demanded. "There is no doubt that all the inhabitants of the unseen world are greatly interested in human agriculture," notes anthropologist Verrier Elwin.[22] Sacrifice is an excess of domestication, involving domesticated animals and occurring only in agricultural societies. Ritual killing, including human sacrifice, is unknown in non-domesticated cultures.[23]

Corn in the Americas tells a parallel story. An abrupt increase in corn agriculture brought with it the rapid elaboration of hierarchy and militarization in large parts of both continents.[24] One instance among many is the northward intrusion of the Hohokams against the indigenous Ootams[25] of southern Arizona, introducing agriculture and organized warfare. By about 1000 A.D. the farming of maize had become dominant throughout the Southwest, complete with year-round ritual observances, priesthoods, social conformity, human sacrifice, and cannibalism.[26] It is hardly an understatement to say, with Kroeber, that with maize agriculture, "all cultural values shifted."[27]

Horses are another instance of the close connection between domestication and war. First domesticated in the Ukraine around 3000 B.C., their objectification fed militarism directly. Almost from the very beginning they served as machines; most importantly, as war machines.[28]

The relatively harmless kinds of intergroup fighting described above gave way to systematic killing as domestication led to increasing competition for land.[29] The drive for fresh land to be exploited is widely accepted as the leading specific cause of war throughout the course of civilization. Once-dominant feelings of gratitude toward a freely giving nature and knowledge of the crucial interdependence of all life are replaced by the ethos of domestication: humans versus the natural world. This enduring power struggle is the template for the wars it constantly engenders. There was awareness of the price exacted by the paradigm of control, as seen in the widespread practice of symbolic regulation or amelioration of domestication of animals in the early Neolithic. But such gestures do not alter the fundamental dynamic at work, any more than they preserve millions of years' worth of gatherer-hunters' practices that balanced population and subsistence.

Agricultural intensification meant more warfare. Submission to this pattern requires that all aspects of society form an integrated whole from which there is little or no escape. With domestication, division of labor now produces full-time specialists in coercion: for example, definitive evidence shows a soldier class established in the Near East by by 4500 B.C. The Jivaro of Amazonia, for millennia a harmonious component of the biotic community, adopted domestication, and "have elaborated blood revenge and warfare to a point where these activities set the tone for the whole society."[30] Organized violence becomes pervasive, mandatory, and normative.

Expressions of power are the essence of civilization, with its core principle of patriarchal rule. It may be that systematic male dominance is a by-product of war. The ritual subordination and devaluation of women is certainly advanced by warrior ideology, which increasingly emphasized "male" activities and downplayed women's roles.

The initiation of boys is a ritual designed to produce a certain type of man, an outcome that is not at all guaranteed by mere biological growth. When group cohesion can no longer be taken for granted, symbolic institutions are required—especially to further compliance with pursuits such as warfare. Lemmonier's judgment is that "male initiations...are connected by their very essence with war."[31]

Polygyny, the practice of one man taking multiple wives, is rare in gatherer-hunter bands, but is the norm for war-making village societies.[32] Once again, domestication is the decisive factor. It is no coincidence that circumcision rituals by the Merida people of Madagascar culminated in aggressive military parades.[33] There have been instances where women not only hunt but also go into combat (e.g. the Amazons of Dahomey; certain groups in Borneo), but it is clear that gender construction has tended toward a masculinist, militarist direction. With state formation, warriorship was a common requirement of citizenship, excluding women from political life.

War is not only ritualistic, usually with many ceremonial features; it is also a very formalized practice. Like ritual itself, war is performed via strictly prescribed movements, gestures, dress, and forms of speech. Soldiers are identical and structured in a standardized display. The formations of organized violence, with their columns and lines, are like agriculture and its rows: files on a grid.[34] Control and discipline are thus served, returning to the theme of ritualized behavior, which is always an increased elaboration of authority.

Exchange between bands in the Paleolithic functioned less as trade (in the economic sense) than as exchange of information. Periodic intergroup gatherings offered marriage opportunities, and insured against resource shortfalls. There was no clear differentiation of social and economic spheres. Similarly, to apply our word "work" is misleading in the absence of production or commodities. While territoriality was part of forager-hunter activity, there is no evidence that it led to war.[35]

Domestication erects the rigid boundaries of surplus and private property, with concomitant possessiveness, enmity, and struggle for ownership. Even conscious mechanisms aimed at mitigating the new realities cannot remove their ever-present, dynamic force. In *The Gift*, Mauss portrayed exchange as peacefully resolved war, and war as the result of unsuccessful transactions; he saw the potlatch as a sort of sub-limated warfare.[36]

Before domestication, boundaries were fluid. The freedom to leave one band for another was an integral part of forager life. The more or less forced integration demanded by complex societies provided a staging ground conducive to organized violence. IN some places, chiefdoms arose from the suppression of smaller communities' inde-pendence. Proto-political centralization was at times pushed forward in the Americas by tribes desperately trying to confederate to fight European invaders.

Ancient civilizations spread as a result of war, and it can be said that warfare is both a cause of statehood, and its result.

Not much has changed since war was first instituted, rooted in ritual and given full-growth potential by domestication. Marshall Sahlins first pointed out that increased work follows developments in symbolic culture. It's also the case that culture begets war, despite claims to the contrary. After all, the impersonal character of civilization grows with the ascendance of the symbolic. Symbols (e.g. national flags) allow our species to dehumanize our fellow-humans, thus enabling systematic intra-species carnage.

4.

The Iron Grip of Civilization: the Axial Age

Civilization is control and very largely a process of the extension of control. This dynamic exists on multiple levels and has produced a few key transition points of fundamental importance.

The Neolithic Revolution of domestication, which established civilization, involved a reorientation of the human mentality. Jacques Cauvin called this level of the initiation of social control "a sort of revolution of symbolism."[1] But this victory of domination proved to be incomplete, its foundations in need of some further shoring up and restructuring. The first major civilizations and empires, in Egypt, China, and Mesopotamia, remained grounded in the consciousness of tribal cultures. Domestication had certainly prevailed—without it, no civilization exists—but the newly dominant perspectives were still intimately related to natural and cosmological cycles. Their total symbolic expressiveness was not yet fully commensurate with the demands of the Iron Age, in the first millennium B.C.

Karl Jaspers identified a turning point for human resymbolization, the "Axial Age",[2] as having occurred between 800 and 200 B.C. in the three major realms of civilization: the Near East (including Greece), India, and China. Jaspers singled out such sixth century prophets and spiritual figures as Zoroaster in Persia, Deutero-Isaiah among the Hebrews, Heraclitus and Pythagoras in Greece, the Buddha in India, and Confucius in China. These individuals simultaneously—but independently—made indelible contributions to post-Neolithic consciousness and to the birth of the world religions.[3] In astonishingly parallel developments, a decisive change was wrought by which civilization established a deeper hold on the human spirit, world-wide.

Internal developments within each of these respective societies broke the relative quiescence of earlier Bronze Age cultures. Wrenching change and new demands on the original patterns were in evidence in many regions. The world's urban population, for example, nearly doubled in the years 600 to 450 B.C.[4] A universal transformation was needed—and effected—providing the "spiritual foundations of humanity" that are still with us today.[5] The individual was fast becoming dwarfed by civilization's quickening Iron Age pace. The accelerating work of domestication demanded a recalibration of consciousness, as human scale and wholeness were left behind. Whereas in the earlier Mesopotamian civilizations, for example, deities were more closely identified with various forces of nature, now society at large grew more differentiated and the separation deepened between the natural and the supernatural. Natural processes were still present, of course, but increasing social and economic tensions strained their integrity as wellsprings of meaning.

The Neolithic era—and even the Bronze Age—had not seen the complete overturning of a nature-culture equilibrium. Before the Axial Age, objects were described linguistically in terms of their activities. Beginning with the Axial Age, the stress is on the static qualities of objects, omitting references to organic processes. In other words, a reification took place, in which outlooks (e.g. ethics) turned away from situation-related discourse to a more abstract, out-of-context orientation. In Henry Bamford Parkes' phrase, the new faiths affirmed "a human rather than a tribalistic view of life."[6]

The whole heritage of sacred places, tribal polytheism, and reverence for the earth-centered was broken, its rituals and sacrifices suddenly out of date. Synonymous with the rise of "higher" civilizations and world religions, a sense of system appeared, and the need for codification became predominant.[7] In the words of Spengler: "the whole world a dynamic system, exact, mathematically disposed, capable down to its first causes of being experimentally probed and numerically fixed so that man can dominate it...."[8] A common aspect of the new reformulation was the ascendance of the single universal deity, who required moral perfection rather than the earlier ceremonies. Increased control of nature and society was bound to evolve toward increased inner control.

Pre-Axial, "animistic" humanity was sustained not only by a less totalizing repression, but also by a surviving sense of union with natural reality. The new religions tended to sever bonds with the manifold, profane world, placing closure on it over and against the supernatural and unnatural.

This involved (and still involves) what Mircea Eliade called "cosmicizing," the passage from a situational, conditional plane to an "unconditioned mode of being."[9] A Buddhist image represents "breaking through the roof"; that is, transcending the mundane realm and entering a trans-human reality.[10] The new, typically monotheistic religions clearly viewed this transcendance as a unity, beyond any particularity of existence. Superpersonal authority or agency, "the most culturally recurrent, cognitively relevant, and evolutionarily compelling concept in religion",[11] was needed to cope with the growing inability of political and religious authority to adequately contain Iron Age disaffection.

A direct, personal relationship with ultimate spiritual reality was a phenomenon that testified to the breakdown of community. The development of individual religious identity, as distinct from one's place in the tribe and in the natural world, was characteristic of Axial consciousness. The personalizing of a spiritual journey and a distancing from the earth shaped human societies in turn. These innovations denied and suppressed indigenous traditions, while fostering the implicit illusion of escaping civilization. Inner transformation and its "way up" was spirit divorced from body, nirvana separate from samsara. Yogic withdrawal, life-denying asceticism, etc. were deeply dualistic, almost without exception.

All this was taking place in the context of an unprecedented level of rationalization and control of daily life in many places, especially by about 500 B.C. S.N. Eisenstadt referred to a resultant "rebellion

against the constraints of division of labor, authority, hierarchy, and...
the structuring of the time dimension. . . "[12] The Axial religions formed
during a period of social disintegration, when long-standing sources
of satisfaction and security were being undermined, and the earlier
relative autonomy of tribes and villages was breaking down. The overall
outcomes were a great strengthening of technological systems, and an
almost simultaneous rise of mighty empires in China (Tsin Shi hwang-
ti), India (Maurya dynasty), and the West (the Hellenistic empires and,
slightly later, the *Imperium Romanum*).

Domestication/civilization set this trajectory in motion by its very
nature, giving birth to technology as domination of nature, and systems based on division of labor. There was mining before 3000 B.C. in
Sinai (early Bronze Age), and a surge in the progress of metallurgical
technology during the third millennium. These innovations coincided
with the emergence of true states, and with the invention of writing.
Naming the stages of cultural development by reference to metals is apt
testimony to their central role. Metallurgy has long stimulated all other
productive activities. By 800 B.C. at the latest, the Iron Age had fully
arrived in the West, with mass production of standardized goods.

Massification of society tended to become the norm, based on specialization. For example, Bronze Age smiths had prospected, mined,
and smelted the ores and then worked and alloyed the metals. Gradually, each of these processes became the purview of corresponding
specialists, eroding autonomy and self-sufficiency. With respect to pottery, a common domestic skill was taken over by professionals.[13] Bread
now came more often from bakeries than from the household. It is
no accident that the Iron Age and the Axial Age commence at almost
exactly the same time, c. 800 B.C. The turbulence and upheavals in the
actual world find new consolations and compensations in the spiritual
realm—new symbolic forms for further fractioning societies. [14]

In Homer's Odyssey (8th century B.C.), the technologically
backward Cyclops have surprisingly easy lives compared to people
in Iron Age Greece of that time, when the beginnings of a factory
system were already in place. Development of steel plows and

weapons accelerated the destruction of nature (erosion, deforestation, etc.) and ruinous warfare.

In Persia, oil was already being refined, if not drilled. There the seer Zoroaster (aka Zarathustra) emerged, providing such potent concepts as immortality, the Last Judgment, and the Holy Spirit (which were quickly incorporated into Judaism). The dualism of the divine Ahura Mazda's struggle against evil was paramount theologically, in a religious system intimately tied to the needs of the state. In fact, the Persian legal system of the Achaemenian period (558-350 B.C.) was virtually synonymous with Zoroastrianism, and the latter in fact quickly became the state religion. According to Harle, Zoroastrianism was "born to serve the demand for social order in a rapidly changing and expanding society." [15]

Zoroastrian monotheism was not only a definitive turning away from animism and the old gods, but also a marked elevation of the categories of good and evil as universals and ruling concepts. Both of these characteristics were Axial Age essentials. Spengler regarded Zarathustra as a "traveling companion of the prophets of Israel", who also steered popular belief away from the web of pantheistic, localist, nature-oriented rites and outlooks. [16]

The Hebrew-Judaic tradition was undergoing a similar change, especially during the same sixth century heart of the Axial Age. The eastern Mediterranean, and Israel in particular, was experiencing a surge of Iron Age urbanization. The social order was under considerable strain in the context of a national need for identity and coherence, especially in the face of more powerful, empire-building neighbors. The Israelites spent two-thirds of the sixth century as captives of the Babylonians.

Yahweh rose from local fertility god to monotheist status in a manner commensurate with the requirements of a beleaguered and threatened people. His grandeur, and the universality of his field of relevance, paralleled the Hebrews' desire for strength in a hostile world.[17] In the eighth century B.C., Amos had announced this vision as a de-ritualizing, transcendentalizing spiritual direction. Jewish uniqueness thus unfolded against the backdrop of radical, unitary divinity.

The "new man" of Ezekiel (early sixth century B.C.) was part of a new supernatural dimension that, again, took its bearings from an unstable time. As Jacob Neusner pointed out, by the sixth century B.C.—at the very latest—the economy was no longer grounded in subsistence or self-sufficiency.[18] The role of the household had been greatly diminished by division of labor and the massifying market. An omnipotent god demanding absolute submission reflected rulers' aspirations for top-down, stabilizing authority. Yahweh, like Zeus, was originally a nature god, albeit connected to domestication. His rule came to hold sway over the moral and civic order, anchored by the rule of kings. The positive, redemptive role of suffering emerged here, unsurprisingly, along with refined political domination. Deutero-Isaiah (Second Isaiah), greatest of the Hebrew prophets of the Axial Age, created a royal ideology in the sixth century B.C.[19] He announced that the very essence of the Covenant with God was embodied in the king himself—that the king *was* the Covenant.[20] The force of this announcement derived from universal cosmic law, beyond any sense perception or earthly parallel; natural phenomena were only its expressions, wrought in an infinity unknowable by mortals.

In pre-Socratic Greece, especially by the time of Pythagoras and Heraclitus in the sixth century B.C., tribal communities were facing disintegration, while new collectivities and institutional complexes were under construction. The silver mines of Laurium were being worked by thousands of slaves. An "advanced manufacturing technology"[21] in large urban workshops often displayed a high degree of division of labor. "Pottery in Athens was made in factories which might employ, under the master-potter, as many as seventy men."[22] Strikes and slave uprisings were not uncommon,[23] while home industries and small-scale cultivators struggled to compete against the new massification. Social frictions found expression, as always, in competing world views.

Hesiod (8th century B.C.) belonged to a tradition of Golden Age proponents, who celebrated an original, uncorrupted humanity. They saw in the Iron Age a further debasing movement away from those origins. Xenophanes (6th century), to the contrary, unequivocally proclaimed that newer was better, echoing Jewish prophets of the Axial Age who had contributed significantly to progressive thinking. He went so far as to see in the forward movement of civilization the origin of all values, glorying in urbanization and increasingly complex technological systems.[24] Xenophanes was the first to proclaim belief in progress.[25] Although the Cynics held out in favor of an earlier vitality and independence, the new creed gained ground. The Sophists upheld its standards, and after 500 B.C., widespread embrace of higher civilization swamped the earlier longing for a primordial, unalienated world.

The transcendentalizing foundation for this shift can be read in an accelerating distancing of people from the land that had been taking place on multiple levels. A land-based pluralism of small producers, with polytheistic attachments to local custom, was transformed by urban growth and stratification, and the detached perspective that suits them. Plato's *Republic* (c. 400 B.C.) is a chilling, disembodied artifact of the rising tendency toward transformation of thought and society along standardized, isolating lines. This model of society was a contrived imposition of the new authoritarianism, utterly removed from the surviving richness that civilization had thus far continued to coexist with.

Social existence intruded to the furthest reaches of consciousness, and the two schema, Iron Age and Axial Age, also overlapped and interacted in India. The period from 1000 to 600 B.C. marked the early Iron Age transition from a socio-economic-cultural mode that was tribal/pastoral, to that of settled/agrarian. The reign of surplus and sedentism was greatly hastened and extended by full-fledged iron and steel plow-based cultivation. Mines and early factories in India also centered on iron technology, and helped push forward the homogenization of cultures in the Mauryan state of this period. New surges of domestication (e.g. horses), urbanization, large estates, and wage labor took place in the Ganges valley, as "tribal egalitarianism," in Romila Thapar's words, surrendered to the newly evolving system by 500 B.C.[26]

This was also roughly the time of Gautama Buddha. Buddhism's origins and role with respect to the spread of Iron Age society can readily be traced.[27] Canonical scriptures refer to early Buddhist teachers as consultants to the rulers of Indian states, a testimony to Buddhism's direct usefulness to the new urban order in a time of great flux. Various commentators have seen the Buddhist reformulation of the premises of Hinduism as an ideology that originated to serve the needs of a challenged, emerging structure.[28] The early supporters, it is clear, were largely members of the urban and rural elites.[29]

For the Buddha—and for the other Axial prophets in general—the personal took precedence over the social. He was the detached observer, seeking freedom from the world, who mainly accepted a very narrow sphere as locus of attention and responsibility. This amounts to a fatalism that founded Buddhism upon suffering as a prime fact, a condition of life that must be accepted. The message of *dukkha* (suffering) expresses the ultimate incapacity of the human condition to include happiness.

Yet Buddhism promised a way out of social dislocation and malaise,[30] through its focus on individual salvation. The goal is "extinguishedness" or Nirvana, the suppression of interest in the world by those disenchanted with it. Similarly, Buddha's presentation of the "cosmic process" was stripped of all earthly processes, human and non-human. While criticizing the caste system and hereditary priesthoods, he took no active role in opposing them. Buddhism was highly adaptive regarding changing social situations, and so was useful to the ruling classes.

Buddhism became another world religion, with global outreach and distinctive superhuman beings to whom prayers are directed. By around 250 B.C. Buddha had become the familiar seated god-figure and Buddhism the official religion of India, as decreed by Asoka, last of the Mauryan dynasty.

The Iron Age came to China slightly later than to India; industrial production of cast iron was widespread by the 4th century B.C. Earlier, Bronze Age polytheism resembled that found elsewhere, complete with a variety of spirits, nature and fertility festivals, etc., corresponding to less specialized, smaller-scale modes of livelihood. The Zhou dynasty had been gradually falling apart since the 8th century; continuous wars and power struggles intensified into the period of the Warring States (482-221 B.C.). Thus the indigenous spiritual traditions, including shamanism and local nature cults, were overtaken by a context of severe technological and political change.

Taoism was a part of this age of upheaval, offering a path of detachment and otherworldliness, while preserving strands of animist spiritual tradition. In fact, early Taoism was an activist religion, with some of its "legendary rebels" engaged in resistance to the new stratifying trends, in favor of re-establishing a class-less Golden Age. [31]

The primitivist theme is evident in the *Chuang Tzu* and survives in the *Tao Te Ching*, key text of Taoism's most prominent voice, Lao-tse (6th century B.C.). An emphasis on simplicity and an anti-state outlook put Taoism on a collision course with the demands of higher civilization in China. Once again, the 500s B.C. were a pivotal time frame, and the opposed messages of Lao-tse and Confucius were typical of Axis Age alternatives.

In contast to Lao-tse, his virtual opposite, Confucius (557-479 B.C.), embraced the state and the New World Order. Instead of a longing for the virtuous time of the "noble savage", before class divisions and division of labor, the Confucian doctrine combined cultural progressivism with the abandonment of connections with nature. No ban was placed on the gods of mountains and winds, ancestral spirits, and the like; but they were no longer judged to be central, or even important.

Confucianism was an explicit adjustment to the new realities, aligning itself with power in a more hands-on, less transcendent way than some other Axial Age spiritualisms. For Confucius, transcendence was mainly inward; he stressed an ethical stringency in service to authority. In this way, a further civilizational colonization was effected, at the level of the individual personality. Internalization of a rigid ruling edifice, minus theology but disciplined by an elaborate code of behavior, was the Confucian way that reigned in China for two thousand years.

These extremely cursory snapshots of Axial Age societies may serve to at least introduce some context to Jaspers' formulation of a global spiritual "breakthrough". The mounting conflict between culture and nature, the growing tensions in human existence, were resolved in favor of civilization, bringing it to a new level of domination. The yoke of domestication was modernized and fitted anew, more tightly than before. The spiritual realm was decisively circumscribed, with earlier, earth-based creeds rendered obsolete. Civilization's original victory over freedom and health was renewed and expanded, with so much sacrificed in the updating process.

The whole ground of spiritual practice was altered to fit the new requirements of mass civilization. The Axial Age religions offered "salvation", at the price of freedom, self-sufficiency, and much of what was left of face-to-face community. Under the old order, the authorities had to use coercion and bribery to control their subjects. Henceforth they could operate more freely within the conquered terrain of service and worship.

The gods were created, in the first place, out of the deepest longings of people who were being steadily deprived of their own authentic powers and autonomy. But even though the way out of progressive debasement was barred by the Axial Age shift, civilization has never been wholeheartedly accepted; and most people have never wholly identified with the "spiritualized" self. How could these ideas be fully embraced, predicated as they were on a mammoth defeat? For Spengler, the Axial Age people who took up these new religions were "tired megalopolitans".[32] Today's faithful, too, may be tired megalopolitans—all too often still spellbound, after all these years, by ideologies of sacrifice, suffering, and redemption.

The renunciations have been legion. Buddhism was founded, for example, by a man who abandoned his wife and newborn child as obstacles to his spiritual progress. Jesus, a few centuries later, exhorted his followers to make similar "sacrifices".

Today's reality of unfolding disaster has a lot to do with the relationship between religion and politics—and more fundamentally, with accepting civilization's trajectory as inevitable. It was the sense of the "unavoidable" that drove people of the sixth century B.C. to the false solutions of Axial Age religiosity; today, our sense of inevitability renders people helpless in the face of ruin, on all fronts. 2500 years is long enough for us to have learned that escape from community, and from the earth, is not a solution, but a root cause of our troubles.

Authentic spirituality is so importantly a function of our connection with the earth. To reclaim the former, we must regain the latter. That so very much stands in our way is the measure of how bereft we have become. Do we have the imagination, strength, and determination to recover the wholeness that was once our human birthright?

5

Alone Together:
the City and its Inmates

The proportion of humanity living in cities has been growing exponentially, along with industrialization. The megalopolis is the latest form of urban "habitat", increasingly interposing itself between human life and the biosphere.

The city is also a barrier between its inmates, a world of strangers. In fact, all cities in world history were founded by strangers and outsiders, settled together in unique, previously unfamiliar environments.

It is the dominant culture at its center, its height, its most dominant. Joseph Grange is, sadly, basically correct in saying that it is "par excellence, the place where human values come to their most concrete expression."[1] (If one pardons the pun, also sadly apt.) Of course, the word "human" receives its fully deformed meaning in the urban context, especially that of today. Everyone can see the modern "flatscape", in Norberg-Schulz's terse term (1969), the Nothing Zones of placelessness where localism and variety are steadily being diminished, if not eradicated.[2] The supermarket, the mall, the airport lounge are everywhere the same, just as office, school, apartment block, hospital, and prison are scarcely distinguishable one from another, in our own cities.[3]

The mega-cities have more in common with each other than with any other social organisms. Their citizens tend to dress the same and otherwise consume the same global culture, under a steadily more comprehensive surveillance gaze. This is the opposite of living in a particular place on the earth, with respect for its uniqueness. These days, all space is becoming urban space; there is not a spot on the planet that couldn't become at least virtually urban upon the turn of a satellite. We have been trained and equipped to mold space as if it were an object. Such an education is mandated in this Digital Age, dominated by cities and metro regions to an extent unprecedented in history.

How has this come to pass? As Weber put it, "one may find anything or everything in the city texts except the informing principle that creates the city itself."[4] But it is clear what the fundamental mechanism/ dynamic/ "principle" is and always has been. As Weber continued: "Every device in the city facilitating trade and industry prepares the way for further division of labor and further specialization of tasks."[5] Further massification, standardization, equivalence.

As tools became systems of technology—that is, as social complexity developed—the city appeared. The city-machine was the earliest and biggest technological phenomenon, the culmination of the division of labor. Or as Lewis Mumford characterized it, "the mark of the city is its purposive social complexity."[6] The two modes in this context are the same. Cities are the most complex artifacts ever contrived, just as urbanization is one of the prime measures of development.

The coming world-city perfects its war on nature, obliterating it in favor of the artificial, and reducing the countryside to mere "environs" that conform to urban priorities. All cities are antithetical to the land.

Certeau's "Walking in the City" has rather an eerie quality, given its subject and the fact that it was written in 2000. Certeau saw the World Trade Center as "the most monumental figure" of Western urbanism and felt that "to be lifted to [its] summit is to be carried away by the city's hold."[7] The viability of the city has entered its inevitable stage of being doubted, accompanied by an anxiety heightened—but not created—on 9/11. The deep ambivalence about urban life, felt throughout civilization's reign, has become much more pronounced.

Domestication made civilization possible, and intensified domestication brought forth urban culture. Primary horticultural communities—settlements and villages—were superseded by cities as massified agriculture took hold. One enduring marker of this shift is megalithic monumentality. In early Neolithic monuments all the qualities of the city are found: sedentism, permanence, density, a visible announcement of the triumphal march of farming over foraging. The city's spectacular centralization is a major turning point in human cultural evolution, the arrival of civilization in its full, definitive sense.

There have been civilizations without cities (e.g. the early Maya civilization), but not many. More often they are a key feature and develop with a relatively sudden force, as if the energy repressed by domestication must burst forth to a new level of its control logic. The urban explosion does not escape some bad reviews, however. In the Hebrew tradition, it was Cain, murderer of Abel, who founded the first city. Similarly, such urban references as Babylon, the Tower of Babel, and Sodom and Gomorrah are wholly negative. A deep ambivalence about cities is, in fact, a constant of civilization.

By about 4000 BC the first cities appeared in Mesopotamia and Egypt, when political means were devised to channel the surpluses created by a new agricultural ethos into the hands of a ruling minority. This development required economic input from wider and wider areas of production; large-scale, centralized, bureaucratic institutions were not long in coming. Villages were pulled into increasingly specialized maximization strategies to produce bigger surpluses flowing to the cities. Greater grain production, for example, could only be achieved with additional work and more coercion. Resistance occurred within this well-known framework, as the more primitive farming communities were forcibly converted into administered towns, such as Nineveh. Nomadic peoples of Sinai refused to mine copper for the Egyptian rulers, to cite another instance.[8] Small-holders were forced off the land into cities; this displacement is a basic part of a familiar pattern that continues today.

Urban reality is primarily about trade and commerce, with a nearly total dependence on support from external areas for continued existence. To guarantee such an artificial subsistence, city fathers turn inevitably to war, that chronic civilizational staple. "Conquest abroad and repression at home," in Stanley Diamond's words, is a defining characterization of cities from their very origins.[9] The early Sumerian city-states, for example, were constantly at war. The struggle for stability of urban market economies was an unremitting matter of survival. Armies and warfare were cardinal necessities, especially given the built-in expansionist character of the urban dynamic. Uruk, the biggest Mesopotamian city of its time (ca 2700 BC), boasted a double-ring wall six miles long, fortified by 900 towers. From this early period through the Middle Ages, virtually all cities were fortified garrisons. Julius Caesar used the word *oppidum* (garrison) to denote every town in Gaul.

The first urban centers also consistently reveal a strong ceremonial orientation. The movement away from an immanent, earth-based spirituality to emphasis on sacred or supernatural spaces receives a further deformation with literally awe-inspiring, mighty urban temples and tombs. The elevation of a society's gods corresponded to the increasing complexity and stratification of its social structure. Religious monumentality, by the way, was not only an obedience-inducing tactic by those in authority; it was also a fundamental vehicle for the spread of domestication.[10]

But the real rise to dominance began not only with intensified agriculture—and the appearance of writing systems, as Childe, Levi-Strauss and others have noted—but with metallurgy. Succeeding civilization's initial Neolithic stage, the Bronze Age and even more so, the Iron Age brought urbanization into its full centrality. According to Toynbee, "If the increase in the size of cities in the course of history is presented visually in the form of a curve, this curve will be found to have the same configuration as a curve presenting the increase in the potency of technology."[11] And with the increasingly urbanized character of social life, the city can be seen as a container. Cities, like the factories that are already present, rely on containment. Cities and factories are never at base freely chosen by the people inside them; domination keeps them there. Aristophanes put it well in his 414 BC creation, *The Birds*: "A city must rise, to house all birds; then you must fence in the air, the sky, the earth, and must surround it by walls, like Babylon."

States as we know them already existed by th'
ful cities emerged as capitals, the loci of state ∙
nation has always flowed from these urban cenu.
peasants leave behind one known and hated servitude for new, _
undisclosed forms of bondage and suffering. The city, already a site of
local power and war, is an incubator of infectious diseases, including
plague, and of course greatly magnifies the impacts of fire, earthquake,
and other dangers.

For thousands of generations humans rose at daybreak and slept
after the sun went down, basking in the glories of sunrise, sunset,
and starry skies. Half a millennium ago, city bells and clocks announced an increasingly ordered and regulated daily life, the reign
of urban timekeeping. With modernity, lived time disappears; time
becomes a resource, an objectified materiality. Measured, reified time
isolates the individual in the force-field of deepening division and separation, ever diminishing wholeness. Contact with the earth ebbs, as
urbanization grows; and as Hogarth depicted in his mid-18th century
images of London, physical contact among people lessens dramatically.
At this time Nicolas Chamfort declared, "Paris is a city of gaieties and
pleasures, where four-fifths of the inhabitants die of grief."[12] In Emile
(1762), Rousseau put it more personally; "Adieu, Paris. We are seeking
love, happiness, innocence. We shall never be far enough away from
you."[13] The pervasive weight of urban existence penetrated even the
most outwardly vital political phenomena, including the French Revolution. Crowds in revolutionary Paris often seemed strangely apathetic,
prompting Richard Sennett to detect there the first pronounced modern signs of urban passivity.[14]

In the following century Engels, in contrary fashion, decided
that it is in the city that the proletariat achieves its "fullest classic
perfection."[15] But Tocqueville had already seen how individuals in
cities feel "strangers to the destinies of each other."[16] Later in the 19th
century, Durkheim noted that suicide and insanity increase with modern urbanization. In fact, a sense of dependence, loneliness, and every
kind of emotional disturbance are generated, giving rise to Benja-

.s perception that "Fear, revulsion, and horror were the emotions ,hich the big-city crowd aroused in those who first observed it."[17] The technological developments in the areas of sewage and other sanitation challenges, while required in burgeoning metropoles, also enable urbanization and its further growth. Life in cities is only possible with such continual technological supports.

By 1900, Georg Simmel understood how living in cities brings about not only loneliness, but also the reserve or emotional numbness that exacerbates it. As Simmel saw, this is very closely analogous to the effects of industrial life in general: "Punctuality, calculability, exactness are forced upon life by the complexity and extension of metropolitan existence."[18] The urban languor and impotence expressed in T.S. Eliot's early poetry, for example, helps fill in this picture of reduced life.

The term "suburb" was used from Shakespeare and Milton onwards in very much the modern sense, but it was not until the onslaught of industrialization that the suburban phenomenon truly emerged. Thus residential development appeared on the outskirts of America's biggest cities between 1815 and 1860. Marx referred to capitalism as "the urbanization of the countryside"[19] ; suburbanization really hit its stride, in its contemporary meaning, just after World War II. Refined mass production techniques created a physical conformity to match and magnify social conformity.[20] Depthless, homogenized, a hothouse of consumerism fenced in by strip malls and freeways, the suburb is the further degraded outcome of the city. As such, the differences between urban and suburban should not be exaggerated or seen as qualitative. Withdrawal, facilitated by an array of high-tech devices—iPods, cell phones, etc.—is now the order of the day, a very telling phenomenon.[21]

Civilization, as is clear from the word's original Latin meaning, is what goes on in cities.[22] More than half of the world's population now lives in cities, McDonaldizing non-places like Kuala Lumpur and Singapore that have so resolutely turned their backs on their own rich contexts. The urbanizing imperative is an ongoing characteristic of civilization.

A certain perverse allure still obtains for some, and it has become so hard to escape the urban influence zone anyway. There is still a flicker of hope for community, or at least for diversion, in the metropolis. And some of us remain there in order not to lose contact with what we feel compelled to understand, so we can bring it to an end. Certainly, there are those who struggle to humanize the city, to develop public gardens and other amenities, but cities remain what they have always been. Most of their inhabitants simply accept the urban reality and try to adjust to it, with the same outward passivity they express toward the enveloping techno-world.

Some try always to reform the unreformable. Let's have "a new modernity", "a new attitude about technology", etc. etc. Julia Kristeva calls for "a cosmopolitanism of a new sort..."[23] Such orientations reveal, among other things, the conviction that what are widely considered essentials of social life will always be with us. Max Weber judged modernity and bureaucratic rationality to be "escape-proof", while Toynbee saw the Ecumenopolis, as he called the stage of gigantism succeeding the stage of the megalopolis, "inevitable".[24] Ellul referred to urbanization as that "which can only be accepted."[25]

However, given today's urban reality, and how and why cities came to be in the first place and continue to exist, what James Baldwin said of the ghetto fully applies to the city: "[It] can be improved in one way only: out of existence."[26] There is a strong consensus among urban theorists, by the way, that "cities are newly divided and polarized."[27] That the poor and the indigenous must be urbanized is another primary facet of colonialist-imperialist ideology.

The original monumentalism is still present and underlined in today's city, with the same dwarfing and disempowering of the individual. Human scale is obliterated by high-rises, sensory deprivation deepens, and inhabitants are assailed by monotony, noise, and other pollutants. The cyberspace world is itself an urban environment, accelerating the radical decline of physical presence and connection. Urban space is the always advancing (vertically and horizontally) symbol of the defeat of nature and the death of community. What John Habberton wrote in 1889 could not be more valid now: "A great city is a great sore—a sore which can never be cured."[28] Or as Kai W. Lee replied to the question whether a transition to sustainable cities is imaginable: "The answer is no."[29]

Copán, Palenque, and Tikal were rich cities of Maya civilization abandoned at their height, between 600 and 900 A.D. With similar examples from various cultures, they point a way forward for us. The literature of urbanism has only grown darker and more dystopian in recent years, as terrorism and collapse cast their shadows on the most untenable products of civilization: the world's cities. Turning from the perpetual servitude and chronic sickness of urban existence, we may draw inspiration from such places as former indigenous settlements on what is now called the Los Angeles River. Places where the sphere of life is rooted in subsisting as fully skilled humans in harmony with the earth.

6

Future Primitive Update

In the past couple of years there have been some very remarkable findings concerning the capacities of early humans.

These discoveries have reinforced and even considerably deepened some aspects of the general paradigm shift underway in recent decades. The work of Thomas Wynn and others has shown that Homo around one million years ago had an intelligence equal to our own. Anthropological orthodoxy now also views Paleolithic humans as essentially peaceful, egalitarian, and healthy, with considerable leisure time and gender equality.

The most recent material has to do with mental achievements and has radical implications similar to those in the other areas of pre-civilized life.

In late August 1999 University of Minnesota and Harvard anthropologists disclosed a narrowing of the size differential between men and women that began about 1.9 million years ago. The key factor was not so much the use of fire, which began then, but cooking of tuberous vegetables. Cooking reduced the need for bigger teeth, which predominated in males, and the sexes began to equalize in size. The fact of cooking, so long ago, is a considerable datum in terms of the capacities of early Homo. An upcoming issue of Current Anthropology will discuss this research in depth.

M. J. Morwood et al., in the March 12, 1998 issue of Nature, revealed evidence that humans used seagoing vessels 800,000 years ago in the western Pacific. The earliest previous evidence for sea crossings dates from about 50,000 years ago. This enormous revision of how long ago humans were able to construct vessels and guide them over miles of ocean actually elicits, according to the authors, a complete reappraisal of the cognitive capacity of early humanity.

In a related vein, a one-million-year-old skull found in Eritrea that possesses Homo sapiens features pushes back such an occurrence by 300,000 to 400,000 years. The September 1998 Discover magazine called this find a "breakthrough in human origins," noting that prior to this discovery, the earliest fossils with H. sapiens features dated to only 700,000 to 600,000 years ago.

The February 27, 1997 issue of Nature recounts the discovery of the world's oldest hunting weapons, a trio of 400,000-year-old wooden spears found in a German coal mine. It is not clear whether this repudiates the prevailing view that Homo engaged almost entirely in foraging or scavenging until about 100,000 years ago, but the find does clearly demonstrate high intelligence. The 6-to-7-foot long spears "required careful planning," utilizing the hardest ends of young spruce trees, with the thickest and heaviest part of the carved shaft about one-third of the distance from the spear point for optimal balance.

What these reports establish is that humans were cooking, traveling over seas, and skillfully making tools at generally much earlier times than previously suspected, and very much prior to any known existence of symbolic culture.

We are trained to equate intelligence with symbolic culture, though clearly this assumption is at variance with the record of human existence. Likewise, we tend to measure intelligence in terms of division of labor and domestication, those benchmarks of basic alienation. We are finding out a bit more about an intelligence that we know lived with nature instead of dominating it, and lived without hierarchy or organized violence. (Head-hunting, cannibalism, slavery, war all appear only with the onset of agriculture.)

On one level or another it seems, humans so very long ago and for so many millennia understood what a good thing they had. Healthy and free, they many have sensed that division of labor erodes wholeness and fragments the individual, leading to social stratification, imbalance, and conflict. They resisted it for more than a million or two million years, succumbing to civilization only quite recently, along with its consolation, symbolic culture.

Beyond Symbolic Thought
A Brief Interview with John Zerzan
By Kevin Tucker, *Species Traitor* #3, Spring 2003.

It seems that no other anarchist has shown as much interest in the concept of symbolic thought as John Zerzan. For the past decades, John has devoted his work to a thorough critique of the totality of civilization, from symbolic thought to the day-to-day misery of this way of life and into the failures of the Left. His essays on the origin of civilization have been collected in Elements of Refusal, Future Primitive and most recently in Running on Emptiness: the Failure of Symbolic Thought. He edited Against Civilization, and is co-editor of Questioning Technology.

Kevin Tucker: How would you distinguish symbolic culture and symbolic thought, and what is their relation to civilization?

John Zerzan: What followed after the species began to symbolize constitutes symbolic culture. This ethos has come to define what thinking is, and the sensual part of experience has to greatly given way to symbolic experience; that is, direct experience is being reduced toward zero point. This narrowed and engineered cultural mode is directly related to civilization, which is the product of continuing control viz. domestication.

Symbolic culture in the forms of art and religion, for example, involve re-presented reality being thus processed as substitute for direct experience. They emerge as societies being to develop inequalities that express themselves in specialized roles and realms of separate authority.

The symbolic may be seen as itself a technology, in that it works upon reality as a force for domination. A similar perspective is Horkheimer and Adorno's "instrumental reason", meaning that civilization comes to infuse or deform rationality itself into patterns of the logic of control.

Freud saw that civilization is that condition necessary for work and culture to triumph; namely, the forcible renunciation of instinctual freedom and Eros. Understood in this sense, it becomes easier to grasp the inner connection between symbolic culture and civilization.

Kevin Tucker: How far back should we be looking with a critique of civilization and why? What is the significance of tracing back so far?

John Zerzan: I don't think it's possible to plumb the depths or origins of civilization without critically examining division of labor or specialization. In the effective power of specialists—possibly the shaman as first case in point—lies the beginning of inequality in human societies. An institution this basic has, of course, been largely overlooked. How can one possibly have modern life without division of labor? But certainly this is just what we are putting into question! Modernity is now seen as increasingly untenable and we are led to wonder at the roots of the extremity of "advanced" society. What propels this trajectory?

Division of labor leads to mass production, even in ancient times, and this requires coordination and justification. Chiefs, bosses, priests flow from this. And the at-first gradual and unnoticed and then rapid development of specialization lays the groundwork for domestication, civilization's defining basis.

Control/ containment takes its next step with private property, but surely the will to dominate animals and plants (domestication) bring civilization rather swiftly, in relative terms. And the nascent elites that are spawned by increasing division of labor provided a stepping-stone to that definitive turn which is domestication.

Kevin Tucker: In your eyes, how does the scientific studies/ research associated with anthropology and archeology weigh compared to what we know now, from ourselves or from the remaining tribes/ bands of various levels of civilized existence?

John Zerzan: I agree with those who say that consulting our own lives is more to the point, more potent than considering the anthropological/ archeological literature. But I think it's also valid to consider evidence from the past that demonstrates an actual state of "natural anarchy" that obtained for such a very long time. Such a picture is an inspiration to me, the realization of the prevalence of non-hierarchical life-ways that constituted the only successful, sustainable adaptation to the world our species has known. Our vision, our critique of civilization is not dependent on such a picture or record, but can draw strength from it.

Most of us are significantly further removed from domesticated existence than any surviving indigenous people. Thus it is important to green anarchy types like myself to learn from them and support their struggles.

Kevin Tucker: How would you know where is something like language took shape? Do you think that the move from language and art necessary brings us to agriculture or is there some middle point of mediation in which we are still embodied by the "other" (wildness)?

John Zerzan: No one knows when language originated. (Speech, that is; we can date written language because of artifactual evidence.) It's one of the most interesting mysteries of all, I'd say. There really are only guesses with some saying it is rather recent (e.g., emerging in the Upper Paleolithic, say, contemporaneously with the earliest cave paintings of about 35,000 years ago) and others figuring that human speech more likely began on the order of a million years back.

If language and art appear more or less together fairly recently virtually on the eve of agriculture then a strong link to domestication is suggested. And obviously if there's a very long time span between their origins, then only art would seem to be linked to domestication.

But it seems quite possible to me that there is a connection – again, at least, in terms of art and agriculture. They are pretty closely related in time, after all.

If speech is very old—and we may never know—then maybe the "middle point of mediation" is that period after speech but before art. That long period when division of labor did not advance and symbolic culture as we know it did not exist.

Kevin Tucker: How do you see the future of civilization and where can a critique of symbolic thought take us?

John Zerzan: Technological civilization is realizing the elimination of the natural world and ever-new depths of individual and social estrangement. It is consuming, impoverishing and destroying its host planet, as everyone can see. It has no future.

A critique of symbolic thought reveals how this malignant virus originated and therefore to what lengths we'll likely have to go to avoid replicating civilization after it falls.

Kevin Tucker: Do you feel that there could be a conscious turn against symbolism and/ or civilization? Or do you feel that the totality of civilization has created dependency relationship that the domesticated will hold strongly to? Do you see anarchy as being brought abut by the domesticated or by those who have turned against their domestication revolting against the agents of civilization?

John Zerzan: There must be a conscious turn against the symbolic and civilization, and I think it has already begun. Antipathy to these dimensions, in fact, is always present, has been present all along, and now it's growing as the generalized crisis deepens.

A "dependency relationship" does obtain, in my opinion, or it could just as easily be called "being held hostage". We will all have to unlearn domestication, and the radical break with domestication in society will most likely occur, I'd say, when it becomes clear that civilization is more of a liability than an asset. When personal immiseration and ecological devastation, for example, reach a certain level and at the same time a viable alternative can be seen as more pleasurable, safe, reasonable.

Part 2:

The Crisis of Civilization

8

Twilight of the Machines

Quite some time ago W.H. Auden summed it up: "The situation of our time surrounds us like a baffling crime." More recently the crisis has been manifesting and deepening in every sphere. Conditions are rapidly worsening and none of the old answers hold up. A friend and neighbor of mine spoke to this with eloquence and understanding: in dealing with others, she counseled, we need to remember that everyone's heart is broken.

Can there really be many left who don't know what direction the world system and this society in particular are taking? Global warming, a function of industrial civilization, will kill the biosphere well before this century is out. Species all over the planet are made extinct at an accelerating rate, dead zones in the oceans grow, the soil and the air are increasingly poisoned, rainforests sacrificed, and all the rest of it.

Children as young as two are on anti-depressants, while emotional disorders among youth have more than doubled in the past 20 years. The teen suicide rate has tripled since the 1970s. A recent study showed that nearly a third of high school students binge drink at least once a month; researchers concluded that "underage drinking has reached epidemic proportions in America."

Meanwhile, most everyone requires some kind of drug just to get through each day, against a backdrop of homicidal outbursts in homes, schools, and workplaces. One of the latest pathological developments—among so many—is parents murdering their children. A panoply of shocking and horrifying phenomena emanate from the disintegrating core of society. We inhabit a landscape of emptiness, grief, stress, boredom, anxiety in which our "human nature" is as steadily degraded as is what is left of the natural world.

The volume of knowledge is reportedly doubling every five years, but in this increasingly technicized, homogenized world an ever-starker reality goes mainly unchallenged, so far. Michel Houellebecq's 1998 novel *Les Particules Elémentaires* (a bestseller in France) captured a joyless, disillusioned modernity in which cloning comes as a deliverance. Civilization itself has proved a failure, and humanity ends up liquidating itself in absolute surrender to domination. How perfectly in tune with the prevailing, completely defeated and cynical postmodern zeitgeist.

Symbolic culture has atrophied our senses, repressed unmediated experience, and brought us, as Freud predicted, to a state of "permanent internal unhappiness." We are debased and impoverished to the point where we are forced to ask why human activity has become so hostile to humanity—not to mention its enmity to other life forms on this planet.

By their very titles, recent books like *All Connected Now: Life in the First Global Civilization and What Will Be: How the New World of Information Will Change Our Lives* express the resignation to an ever more standardized and bereft situation. Such works express the creative exhaustion and moral bankruptcy of the age, in which massive dehumanization and rampant destruction of nature vie for fulfillment of their interrelated projects.

1997-98 saw several months of smoke all across Southeast Asia as four million hectares of forests burned. Four years later, hundreds of fires raged for many weeks across eastern Australia, set by bored teenagers. In the U.S., groundwater and soil pollution levels have risen measurably because of concentrations of anti-depressants in human urine. Alienation in society and the annihilation of plant and animal communities join in a ghastly, interlocked dance of violence against health and life.

Reified existence progressively disables whatever and whoever questions it. How else to account for the stunningly accommodationist nature of postmodernism, allergic to any questioning of the basics of the reigning techno-capitalist malignancy? And yet a questioning is emerging, and is fast taking shape as the deep impetus of a renewed social movement.

As the life-world's vital signs worsen on every level, the best minds should be paying close attention and seeking solutions. Instead, most have found an infinitude of ways to ponder the paralyzing dichotomy of civilization versus nature, unable to reach an increasingly unavoidable conclusion. A few farsighted individuals began the questioning in modern times. Horkheimer came to realize that domination of nature and humans, and the instrumental reason behind that domination, flow from the "deepest layers of civilization." Bataille grasped that "the very movement in which man negates Mother Earth who gave birth to him, opens the path to subjugation."

After about thirty years without social movements, we are seeing a rebirth. Driven and informed by the growing crisis in every sphere, reaching deeper for understanding and critique than did the movement of the 1960s, the new movement is "anarchist," for want of a better term. Ever since the several days' anti-World Trade Organization militancy in the streets of Seattle in November 1999, the orientation of anti-globalization has become steadily more evident. "Anarchism is the dominant perspective within the movement," Barbara Epstein judged in a fall 2001 report. Esther Kaplan observed in February 2002 that "as the months have rolled by since Seattle, more and more activists, with little fanfare, have come to explicitly identify as anarchists, and anarchist-minded collectives are on the rise. . . . The anarchist fringe is fast becoming the movement's center." David Graeber put it even more succinctly: "Anarchism is the heart of the movement, its soul; the source of most of what's new and hopeful about it."

Henry Kissinger referred to the anti-globalization protests of 1999 and 2000 as "early warning signals" of a "potential political weight" in the industrialized countries and the Third World, as a threat to the world system itself. A CIA report that became public in spring 2000, "Global Trends 2015," predicted that the biggest obstacle to globalization in the new millennium would be a possible joining together of the "First World" protest movement with the struggles of indigenous people to maintain their integrity against encroaching capital and technology.

Which introduces a more important question about this movement and its threatening connection to the centuries of struggles against Empire in the not-yet-industrialized world. Namely, if it is increasingly anarchy-oriented, what does this anarchism consist of?

I think it is fairly clear that it is becoming something other than part of the left. Until now, every modern anti-capitalist movement had at its core an acceptance of the expansion of the means of production and the continuing development of technology. Now there is an explicit refusal of this productionist orientation; it is in the ascendant in the new anarchy movement.

This anarcho-primitivist (or simply primitivist) tendency knows that to account for today's grim realities there needs to be a deeper look at institutions once almost universally taken for granted. Despite the postmodern ban on investigation of these institutions' origins, the new outlook brings even division of labor and domestication into question as ultimate root causes of our present extremity of existence. Technology, meaning a system of ever greater division of labor or specialization, is indicted as the motor of ever greater technicization of the life-world. Civilization, which arrives when division of labor reaches the stage that produces domestication, is also now seen as deeply problematic. Whereas the domestication of animals and plants was once assumed as a given, now its logic is brought into focus. To see the meaning of genetic engineering and human cloning, for example, is to grasp them as implicit in the basic move to domination of nature, which is domestication. Though it is apparent that this critical approach raises more questions than it answers, a developing anarchy consciousness that does not aim at definitive answers cannot turn back.

Cannot turn back to the old, failed left, that is. Who doesn't know at this point that something different is urgently needed?

One of the touchstones or inspirations of primitivist anarchy is the paradigm shift in the fields of anthropology and archaeology in recent decades, concerning human social life during "prehistory." Civilization appeared only some 9,000 years ago. Its duration is dwarfed by the thousands of human generations who enjoyed what might be called a state of natural anarchy. The general orthodoxy in the anthropological literature, even including textbooks, portrays life outside of civilization as one of ample leisure time; an egalitarian, food-sharing mode of life; relative autonomy or equality of the sexes; and the absence of organized violence.

Humans used fire to cook fibrous vegetables almost two million years ago, and navigated on the open seas at least 800,000 years ago.. They had an intelligence equal to ours, and enjoyed by far the most successful, non-destructive human adaptation to the natural world that has ever existed. Whereas the textbook question used to be, "Why did it take Homo so long to adopt domestication or agriculture?" now texts ask why they did it at all.

As the negative and even terminal fruits of technology and civilization become ever clearer, the shift to a luddite, anti-civilization politics makes greater sense all the time. It is not very surprising to detect its influence being registered in various circumstances, including that of the massive anti-G8 protests in Genoa, July 2001. 300,000 people took part and $50 million in damage was caused. The Italian minister of the interior blamed the anarchist "black bloc," and its supposed primitivist outlook in particular, for the level of militancy.

How much time do we have to effect what is necessary to save the biosphere and our very humanness? The old approaches are so many discredited efforts to run this world, which is a massified grid of production and estrangement. Green or primitivist anarchy prefers the vista of radically decentralized, face-to-face community, based on what nature can give rather than on how complete domination of nature can become. Our vision runs directly counter to the dominant trajectory of technology and capital, for the most obvious of reasons.

The left has failed monumentally, in terms of the individual and in terms of nature. Meanwhile, the distance between the left and the new anarchy movement keeps widening. Pierre Bourdieu and Richard Rorty, for instance, long absurdly for a renewed connection between intellectuals and unions, as if this chimera would somehow change anything on a basic level. Jurgen Habermas' *Between Facts and Norms* is an apologetic for things as they are, blind to the real colonization of modern life, and even more uncritical and affirmative than his previous works. Hardt and Negri speak to the choice involved rather directly: "We would be anarchists if we were not to speak... from the standpoint of a materiality constituted in the networks of productive cooperation, in other words, from the perspective of a humanity that is constructed productively.... No, we are not anarchists but communists." Conversely, to further clarify the issue, Jesus Sepulveda observed that "anarchy and indigenous movements fight against the civilized order and its practice of standardization."

Not all anarchists subscribe to the increasing suspicions about technology and civilization. Noam Chomsky and Murray Bookchin, for example, insist on the traditional embrace of progressive development. The marxian heart of anarcho-syndicalism typifies this adherence and is fading away with its leftist relatives.

Marx, who knew so much about the impact of the productive process and its destructive course as division of labor, nonetheless believed (or wanted to believe) that the technological dynamic would undermine capitalism. But "all that is solid" does *not* "melt into air"; rather it becomes more like what it always was. This is as true for civilization as for capitalism.

And civilization now has the form technology gives it, inseparable from the rest of the social order—the world landscape of capital—and embodying and expressing its deepest values. "We have only purely technological conditions left," concluded Heidegger, whose formulation is itself sufficient to expose the myth of technology's "neutrality."

At its origin in division of labor and until now, technology has been an assumption, repressed as an object of attention. At the point when generalized technicization characterizes the world and is the most dominant aspect of modern life, the veil is being lifted. Technology's invasive colonization of everyday life and systematic displacement of the physical environment can no longer be ignored or concealed. A thousand questions push forward.

Health is just one, as we witness the resurgence and multiplication of diseases, increasingly resistant to the industrial medicine that claimed to be erasing them. Antidepressants mask some of the symptoms of rising levels of sorrow, depression, anxiety, and despair, while we are supposed to remain in the dark about the multisensory richness, diversity, and immediacy that technology leaches out of our lives. Cyberspace promises connection, empowerment, variety to people who have never been so isolated, disempowered, and standardized. Each new study confirms that even a few hours' internet use produces the latter effects. Technology has also served to extend the reach of work via the many gadgets, especially cell phones, beepers, and e- mail, that keep millions in harness regardless of time or place.

What is the cultural ethos that has blunted criticism and resistance and, in effect, legitimated the illegitimate? None other than postmodernism, which may have finally reached the nadir of its moral and intellectual bankruptcy.

Seyla Benhabib provides a compelling version of postmodern thought in three theses: "the death of man understood as the death of the autonomous, self-reflective subject, capable of acting on principle; the death of history, understood as the severance of the epistemic interest in history of struggling groups in constructing their past narratives; the death of metaphysics, understood as the impossibility of criticizing or legitimizing institutions, practices, and traditions other than through the immanent appeal to self-legitimation of 'small narratives'." Marshall Berman encapsulates postmodernism as "a philosophy of despair masquerading as radical intellectual chic...the counterpoint to the civilizational collapse going on around us."

Postmodernists champion diversity, difference, and heterogeneity, choosing to see reality as fluid and indeterminate. The actual parallel to this attitude is found in the movement of commodities with brief shelf-lives, circulating meaninglessly in a globalized, fast-food hip consumerism. Postmodernism insists on surface, and is at pains to discredit any notion of authenticity. No deep meanings are accepted; universals of any kind are scorned in favor of a supposed particularity. The meaning of a universal, homogenizing technology, on the other hand, is not only unquestioned, but is embraced. The connection between the imperialism of technology and the loss of meaning in society never dawns on the postmodernists.

Born of the defeat of the movements of the 1960s and grown ever more embarrassingly impoverished during the post-'60s decades of defeat and reaction, postmodernism is the name for prostration before the monstrous facts. Happy to accept the present as one of technonature and technoculture, Donna Haraway epitomizes the postmodern surrender. Technology, it seems, always was; there is no way to stand outside its culture; the "natural" is no more than the pervasive naturalization of culture. In sum, there is no "nature" to defend, "we're all cyborgs." This stance is obviously of benefit in the war against nature; more specifically in the wars against women, indigenous cultures, surviving species, indeed against all of non-engineered life.

For Haraway, technological prosthesis "becomes a fundamental category for understanding our most intimate selves" as we merge with the machine. "Technoscience...[is] unmistakably science for us." Unsurprisingly, she has chided those who would resist genetic engineering, with the reminder that the world is too "unsettled, dirty" for simplistic verdicts about the practices of technoscience. In truth, opposing it is "redactive" and "foolish."

Sadly, there are all too many who follow her path of capitulation to the death-trip we've been forced on. Daniel R. White writes, rather incredibly, of "a postmodern-ecological rubric that steps past the traditional either-or of the Oppressor and the Oppressed." He further muses, echoing Haraway: "We are all becoming cyborgs. What sort of creatures do we want to be? Do we want to be creatures at all? Would machines be better? What kinds of machines might we become?"

Michel Foucault was, of course, a key postmodern figure whose influence has not been liberatory. He ended up losing his way in the area of power, concluding that power is everywhere and nowhere; this argument facilitated the postmodern conceit that opposing oppression is passé. More specifically, Foucault determined that resisting technology is futile, and that human relations are inescapably technological.

The postmodern period, according to Paul Virilio, is "the era of the sudden industrialization of the end, the all-out globalization of the havoc wreaked by progress." We must move past postmodern accommodation and undo this progress.

Civilization is the foundation that decides the rest. As Freud noted, "there are difficulties attaching to the nature of civilization which will not yield to any attempt at reform." "Difficulties" stemming from the origin of civilization as the forced renunciation of Eros and instinctual freedom; "difficulties" that, as he predicted, will produce a state of universal neurosis.

Freud also referred to "the sense of guilt produced by civilization... [which] remains to a large extent unconscious, or appears as sort of malaise, a dissatisfaction." The magnitude of the offense which is civilization explains this great, continuing quantum of guilt, especially since the continual re-enactment of the offense—the curbing of instinctual freedom— is necessary to maintain the coercion and destructiveness that is civilization.

Spengler, Tainter and many others concur that collapse is inherent in civilizations. We may be approaching the collapse of this civilization more quickly than we can grasp, with results even more unimaginable. Along with the rapid degradation of the physical world, are we not seeing a disintegration of the symbol system of Western civilization? So many ways to register the sinking credibility of what is ever more nakedly the direct rule of technology and capital. Weber, for example, identified the disfiguring or marginalization of face-to-face ethical sensibilities as the most significant consequence of modern processes of development.

The list of crimes is virtually endless. The question is whether or not, when the civilization comes down, it will be allowed to recycle into one more variant of the original crime.

The new movement replies in the negative. Primitivists draw strength from their understanding that no matter how bereft our lives have become in the last ten thousand years, for most of our nearly two million years on the planet, human life appears to have been healthy and authentic. We are moving, this anti-authoritarian current, in the direction of primitive naturalism, and against a totality that moves so precisely away from that condition. As Dario Fo put it, "The best thing today is this fantastic breeze and sun—these young people who are organizing themselves across the world." Another Italian voice filled out this sentiment admirably: "And then at bottom, what really is this globalization of which so many speak? Perhaps the process of the expansion of markets toward the exploitation of the poorest countries and of their resources and away from the richer countries? Perhaps the standardization of culture and the diffusion of a dominant model? But then, why not use the term civilization that certainly sounds less menacing but is fitting, without the necessity of a neologism. There is no doubt that the media—and not just the media—have an interest in mixing everything in a vague anti-globalization soup. So it's up to us to bring clarity to things, to make deep critiques and act in consequence." (Terra Salvaggio, July 2000).

It's an all or nothing struggle. Anarchy is just a name for those who embrace its promise of redemption and wholeness, and try to face up to how far we'll need to travel to get there. We humans once had it right, if the anthropologists are to be believed. Now we'll find out if we can get it right again.

Quite possibly our last opportunity as a species.

9

Exiled from Presence

The Unbearable Lightness of Being keeps occurring to me, in a mocking sort of way. Not so much the substance of Kundera's novel, but the title itself. It is the "lightness" of an ever-more-complete disembodiment that is pronounced—and becoming unbearable. A fully technified existence is overtaking us, redefining everything according to the terms of what is not present. It is the triumph of the virtual, the cyber/cyborgian/digital. We exist in the age of floating Data and dispossessed people where the remote reigns, under the surveillance of an Information Technology hypercomplexity.

"We've become weightless, in the bad sense of the word," says a character in Alan Lightman's novel, *Reunion*.[1] Virtual. Almost. Receding. As Stephen Erickson put it, "Philosophy itself surely is at a crossroad, a marker for which is a pervasive kind of emptiness."[2]

A trend toward disembodiment is not only an abiding strain of Western intellectual history; it is an inescapable result of civilization. For at least 10,000 years, the very ground of our shared existence as humans has been shifting, drawing us away from being present to the world, to each other, to ourselves.

Descartes found it problematic that embodiment shapes our understanding of the world. His project was to overcome the limitations of a body that clouds the mind's purchase on the world. The Cartesian outlook is as central to the logic of modern technology as it is to Descartes' position as the founder of modern philosophy.

We now inhabit a world whose condition is not quite fatal, where modernity is completing the abolition of nature. Is there an escape hatch from this destructive course? Who is looking for a way out? The antidote to Descartes—and the terminal progress he helped engender—may become clearer, sooner than we think. The results of not changing everything about human culture loom ahead, increasingly predictable and even obvious. A radical shift is essential to avert disaster.[3] In referring to "the sensory modes and philosophical genius of indigenous peoples," Paul Shepard[4] provided a benchmark reminder that the answers we seek may be very old, and still applicable.

The dominant counter-voices are prevailing, needless to say, and delivering on their Cartesian promises. "Perhaps one day virtual environments, and the synthetic reasoning they allow to emerge, will become the tools we need...in our search for a better destiny for humanity," offers Manuel de Landa.[5] Perhaps, tragically, this desperate, hopeless gambling away of what's left of this planet will only be refused when things get far worse.

In the early to mid-20th century, a few voices rose in opposition to Descartes' cult of disembodiment. For the phenomenologically oriented philosopher Edmund Husserl, bodily perception is the prime starting point for thought. Embodiment, or presence, is foundational and lies beyond the operation of language. Describing the origins of human consciousness, Husserl held that an undivided self preceded humankind's adoption of symbolic thought.

The postmodern outlook began with Heidegger's break with the concept of a foundational presence. Derrida's early, defining texts of the 1950s and 1960s were part of his effort to discredit Husserl's basic premise. Derrida approached the term *presence* as essentially a space outside of representation. In his view, it is an illusion to imagine that such a non-mediated condition has ever existed. Because Husserl questioned whether all experience is necessarily subject to objectification and separation, Derrida had to take him on.

Against Husserl's watchword, "Return to the things themselves,"[6] Derrida opposed his famous dictum that there is nothing outside language or symbolic culture. In his words, "Certainly nothing has preceded this [mediated] situation. Assuredly nothing will suspend it." Postmodernism, the handmaiden of technoculture's malignant spread, must deny that matters were ever outside the realm of estrangement, or ever will be.

Deconstruction is in the end always aimed at presence, and makes one very important point. It exposes the illusion whereby language evokes an object's presence, rather than its absence. But deconstructionists over-shoot the mark when they declare that there is nothing of meaning outside the text.

Derrida begins with the notion that the subject can have no self-presence except by hearing him or herself speak. This view of consciousness fails to acknowledge that thought is not solely linguistic, even in this overly symbolized world (viz. dreaming, listening to and playing music, sports, and countless other examples). For Derrida, "It is difficult to conceive anything at all beyond representation."[8]

In true Cartesian fashion, Derrida flees embodiment and the possibility of a world directly lived. It's telling that he views time as a metaphysical conceptuality, rather than an onerous dimension that can serve as an index of alienation.

With language, the body retreats, effacing itself. When the mouth produces linguistic signs (words), the rest of the body disappears. Presence has an inverse relationship to language; it is most apparent when people are silent. Especially in its written form, language stands separate from the non-reified, processual flow of life. The creator of the Cherokee writing system, Sequoyah, observed that "White men make what they know fast on paper like catching a wild animal and taming it."[9]

In the 18th century, Rousseau attacked representation in the name of unmediated presence. His goal of fully "transparent" relationships implies that representation is the undoing of that which must necessarily be face-to-face. Derrida, on the other hand (echoing Hegel) views the notion of a transparent relationship as mere nostalgic fantasy, a result of the unhappy consciousness of modernity. Not only is immediacy no longer available to us because of complexity and modernity, Derrida holds that it never really existed at all. His *Specters of Marx*[10] underlines Derrida's refusal to question a globalized techno-reality and the "democracy" it requires. Disembodiment and representation are the facts of life; we have no choice but to submit.

Indeed, a visitor from another planet might readily conclude that representation is all there is on earth today. Mediation is steadily burying simple directness. The linearization of symbols as language prefigured a triumphing technology, and a world built of images is now a dominant fact. More and more, the commodity is the same as the sign, in a society dedicated, above all, to consuming. This is the real status of symbolic "meaning" which rules by the principle of equivalence—unlike a gift, which is given without expectation of an equivalent return. Symbolic culture swallows and defines the landscape; no part of it is distinct from the underlying dis-embodying movement that is overtaking what is left of presence. Commodification and aestheticization of the life-world proceed hand in hand. Consuming and ineffective, stylistic gestures prevail.

Instead of being able to "return to the things themselves" à la Husserl, the channeled current of technology carries us further from them. Technology is the shock-troop of modernity, as Paul Piccone put it[11], and on technoculture's horizon the object simply fades out.

Postmodernism functions affirmatively for this advance, complementing its negative role of enforcing a ban on non-symbolic being. Rushing out to embrace the bleakness of techno-cultural existence, postmodern thinkers prepare and support its continuing success. Gregory Ulmer's *Applied Grammatology* celebrates the key Derridean notion of its title as facilitating new "technologies" of writing that are consciously allied with progress. According to Ulmer, grammatology makes possible a "new organization of cultural studies" that is responsive to the prevailing "era of communications technology."[12] *Hypertext: The Convergence of Contemporary Critical Theory and Technology* by George Landow[13] makes perfectly clear, beginning with its title, the subservience of postmodern doctrine to technoculture.

Of course, it is division of labor that drives the growing complexity and all it brings in its wake. From the earliest erosion of human connectedness, immediacy, autonomy, and equality, an ever-intensifying division of labor has driven industrial technics as well as social inequality. The fragmentation of experience advances into all sectors of human life by means of this foundational social institution. Heidegger remained fundamentally ambivalent concerning the essence of technology because he refused to ground his thought in anthropological and historical reality. While recognizing the progressive domination of everything by unfolding technology, he also decided, characteristically, that "human activity can never directly counter this danger." [14]

Thanks to the postmodern know-nothingism and failure of nerve, reality turns out to be not so bad after all. There's nothing one can do about it anyway; Heidegger himself said so. Ellen Mortensen speaks for the general accommodation: "My inclination is to accept the fundamental indeterminacy attached to technology, and to be open to the fact that the way of revealing that technology seems to govern today might possibly change as a result of future radical changes in technology." [15] Could anything be more comforting to the ruling order and its direction?

Postmodernism is predicated on the thesis that the all-enveloping symbolic atmosphere, foundationless and inescapable, is made up of shifting, indeterminate signifiers that can never establish firm meaning. It is in this sense that Timothy Lenoir defends the fusing of life and machine while rebuking critics of high-tech dehumanization: "I wonder where one might locate the moral high ground in order to fashion such a critical framework?" [16] The reigning cultural ethos has explicitly denied the possibility of such ground or stable locus of meaning and value. Criticism is disarmed.

This sensibility, and its intended result, are further along than most people realize. Worth quoting at length, as one example, is Kathleen Woodward's non-atypical postmodern take on the future of human feelings in reference to "the process of technocultural feedback loops generating emotional growth."

> "The emergence of intersubjectivity between the human-
> world and the technological world (represented by replicants

and nonhuman cyborgs) results in a form of intelligence—
emotional intelligence—that is not only resourceful in a
multitude of ways but is also deeply benevolent.... What
is ultimately represented, then, is a system of distributed
emotional intelligence where the human mindbody has
profoundly meaningful ties to the cyberworld, feelings that
are reciprocated. [17]

Jacquère Roseanne Stone looks "forward eagerly to continuing this
high adventure" with technology "as we inexorably become creatures
that we cannot even now imagine," celebrating such joys "at the dawn
of the virtual age,"etc. [18] What is inexorable, like our basic entrapment
in an ever-greater mediation, might as well be rejoiced in, apparently.

But it is obvious that some find all this offensive, even shocking, if
this profoundly unhealthy society hasn't succeeded yet in rendering us
immune to further shock. Mort, May and Williams,[19] for instance, discuss "telemedicine", by which the human touch is completely removed
from so-called healing. They are aware that the soon-to-be virtual clinic
must be integrated into the larger socio-technological ensemble, and
that confidence in this arid project is essential. Their doubts about the
latter are manifest.

It isn't as if there are no choices. After all, choice is mandatory in
the consumer society, just as it is with deconstruction, where the supposedly fluid open-endedness of things requires multiple, continually
revised interpretations. But the choices, so defined, are equally incon-
sequential in either (closely related) domain. The sickness of modernity
will not be cured by more doses of modernity.

Behind the baffling failures of our hollow, distancing life arrange-
ments stands the overall failure of complexity itself, considered from
whatever angle one chooses. A deeo shift must arise, as ever, from doubt
and need. What grows unchecked is a vista that can only give rise to
doubt because it fails to satisfy a single authentic need.

What need could be greater than our hunger for presence, for
that quality that is so primary, whose absence is the measure of our

impoverishment? Adorno dismissed as busywork any philosophy that doesn't risk total failure.[20] It's no wonder that many have concluded that the "end of philosophy" has been reached, when avoidance of the obvious remains the rule—though everything is at stake!

Symbolic culture has always been a mediated or virtual reality, long before what we call "media" existed. With hypertext, hypermedia, and the like, in the era of hypercomplexity, it is easier now to see what the destination of culture has been all along. Not from our beginning, but from its beginning. As the rate of completion of technification accelerates, so does the diet of fantasy and denial grow in importance, consumed by a jaded and enervated humanity.

Who can deny that instead of more retreat from reality, we need a life-centered re-embodiment, a return to groundedness, presence, the face-to-face? The promise of communication is imbued with that dimension. We say we're "in touch with," "in contact with" another. That is, we'd like to be, while directness continues to be systematically drained from even our closest relationships.

Novalis called philosophy a kind of homesickness, reminding us of the lost unity of the world. "Nature is always for us as at the first day," said the phenomenologist Merleau-Ponty[21], offering an inspiring perspective. An understanding of origins, so categorically ruled out by postmodernists, may be among the necessary antidotes to an otherwise terminal condition.

For several decades now, pessimism has been the norm for supposedly critical thinking. Imagine the difference if the inevitability of technological civilization were no longer a given.

10

The Modern Anti-World

There now exists only one civilization, a single global domestication machine. Modernity's continuing efforts to disenchant and instrumentalize the non-cultural natural world have produced a reality in which there is virtually nothing left outside the system. This trajectory was already visible by the time of the first urbanites. Since those Neolithic times we have moved ever closer to the complete de-realization of nature, culminating in a state of world emergency today. Approaching ruin is the commonplace vista, our obvious non-future.

It's hardly necessary to point out that none of the claims of modernity/Enlightenment (regarding freedom, reason, the individual) are valid. Modernity is inherently globalizing, massifying, standardizing. The self-evident conclusion that an indefinite expansion of productive forces will be fatal deals the final blow to belief in progress. As China's industrialization efforts go into hyper-drive, we have another graphic case in point.

Since the Neolithic, there has been a steadily increasing dependence on technology, civilization's material culture. As Horkheimer and Adorno pointed out, the history of civilization is the history of renunciation. One gets less than one puts in. This is the fraud of technoculture, and the hidden core of domestication: the growing impoverishment of self, society, and Earth. Meanwhile, modern subjects hope that somehow the promise of yet more modernity will heal the wounds that afflict them.

A defining feature of the present world is built-in disaster, now announcing itself on a daily basis. But the crisis facing the biosphere is arguably less noticeable and compelling, in the First World at least, than everyday alienation, despair, and entrapment in a routinized, meaningless control grid.

Influence over even the smallest event or circumstance drains steadily away, as global systems of production and exchange destroy local particularity, distinctiveness, and custom. Gone is an earlier pre-eminence of place, increasingly replaced by what Pico Ayer calls "airport culture"—rootless, urban, homogenized. Modernity finds its original basis in colonialism, just as civilization itself is founded on domina-tion—at an ever more basic level. Some would like to forget this pivotal element of conquest, or else "transcend" it, as in Enrique Dussel's facile "new trans-modernity" pseudo-resolution (*The Invention of the Ameri-cas*, 1995). Scott Lash employs somewhat similar sleight-of-hand in *Another Modernity: A Different Rationality* (1999), a feeble nonsense title given his affirmation of the world of technoculture. One more tortuous failure is *Alternative Modernity* (1995), in which Andrew Feenberg sagely observes that "technology is not a particular value one must choose for or against, but a challenge to evolve and multiply worlds without end." The triumphant world of technicized civilization—known to us as modernization, globalization, or capitalism—has nothing to fear from such empty evasiveness.

Paradoxically, most contemporary works of social analysis provide grounds for an indictment of the modern world, yet fail to confront the consequences of the context they develop. David Abrams' *The Spell of the Sensuous* (1995), for example, provides a very critical overview of the roots of the anti-life totality, only to conclude on an absurd note. Ducking the logical conclusion of his entire book (which should be a call to oppose the horrific contours of techno-civilization), Abrams decides that this movement toward the abyss is, after all, earth-based and "organic." Thus "sooner or later [it] must accept the invitation of gravity and settle back into the land." An astoundingly irresponsible way to conclude his analysis.

Richard Stivers has studied the dominant contemporary ethos of loneliness, boredom, mental illness, etc., especially in his *Shades of Lone-liness: Pathologies of Technological Society* (1998). But this work fizzles out into quietism, just as his critique in *Technology as Magic* ends with a similar avoidance: "the struggle is not against technology, which is a simplistic understanding of the problem, but against a technological system that is now our life-milieu."

The Enigma of Health (1996) by Hans Georg Gadamer advises us to bring "the achievements of modern society, with all of its automat- ed, bureaucratic and technological apparatus, back into the service of that fundamental rhythm which sustains the proper order of bodily life". Nine pages earlier, Gadamer observes that it is precisely this apparatus of objectification that produces our "violent estrangement from ourselves."

The list of examples could fill a small library—and the horror show goes on. One datum among thousands is this society's staggering level of dependence on drug technology. Work, sleep, recreation, non-anxi- ety/depression, sexual function, sports performance—what is exempt? Anti-depressant use among preschoolers—*preschoolers*—is surging, for example (*New York Times*, April 2, 2004).

Aside from the double-talk of countless semi-critical "theorists", however, is the simple weight of unapologetic inertia: the count- less voices who counsel that modernity is simply inescapable and we should desist from questioning it. It's clear that there is no escaping modernization anywhere in the world, they say, and that is unalter- able. Such fatalism is well captured by the title of Michel Dertourzos' *What Will Be: How the New World of Information Will Change Our Lives* (1997).

Small wonder that nostalgia is so prevalent, that passionate yearning for all that has been stripped from our lives. Ubiquitous loss mounts, along with protest against our uprootedness, and calls for a return home. As ever, partisans of deepening domestication tell us to abandon our desires and grow up. Norman Jacobson ("Escape from Alienation: Challenges to the Nation-State," *Representations* 84: 2004) warns that nostalgia becomes dangerous, a hazard to the State, if it leaves the world of art or legend. This craven leftist counsels "realism" not fantasies: "Learning to live with alienation is the equivalent in the political sphere of the relinquishment of the security blanket of our infancy."

Civilization, as Freud knew, must be defended against the individual; all of its institutions are part of that defense.

But how do we get out of here—off this death ship? Nostalgia alone is hardly adequate to the project of emancipation. The biggest obstacle to taking the first step is as obvious as it is profound. If understanding comes first, it should be clear that one cannot accept the totality and also formulate an authentic critique and a qualitatively different vision of that totality. This fundamental inconsistency results in the glaring incoherence of some of the works cited above.

I return to Walter Benjamin's striking allegory of the meaning of modernity:

> His face is turned toward the past. Where we perceive a chain of events, he sees one single catastrophe which keeps piling ruin upon ruin and hurls it in front of his feet. The angel would like to stay, awaken the dead and make whole what has been smashed. But a storm is blowing from Paradise; it has got caught in his wings with such violence that the angel can no longer close them. The storm irresistibly propels him into the future to which his back is turned, while the pile of debris before him grows skyward. This storm is what we call progress. (1940)

There was a time when this storm was not raging, when nature was not an adversary to be conquered and tamed into everything that is barren and ersatz. But we've been traveling at increasing speed, with rising gusts of progress at our backs, to even further disenchantment, whose impoverished totality now severely imperils both life and health.

Systematic complexity fragments, colonizes, debases daily life. Division of labor, its motor, diminishes humanness in its very depths, dis-abling and pacifying us. This de-skilling specialization, which gives us the illusion of competence, is a key, enabling predicate of domestication.

Before domestication, Ernest Gellner (*Sword, Plow and Book*, 1989) noted, "there simply was no possibility of a growth in scale and in complexity of the division of labour and social differentiation." Of course, there is still an enforced consensus that a "regression" from civilization would entail too high a cost—bolstered by fictitious scary scenarios, most of them resembling nothing so much as the current products of modernity.

People have begun to interrogate modernity. Already a specter is haunting its now crumbling façade. In the 1980s, Jurgen Habermas feared that the "ideas of antimodernity, together with an additional touch of premodernity," had already attained some popularity. A great tide of such thinking seems all but inevitable, and is beginning to resonate in popular films, novels, music, zines, TV shows, etc.

And it is also a sad fact that accumulated damage has caused a widespread loss of optimism and hope. Refusal to break with the totality crowns and solidifies this suicide-inducing pessimism. Only visions completely undefined by the current reality constitute our first steps to liberation. We cannot allow ourselves to continue to operate on the enemy's terms. (This position may appear extreme; 19th century abolitionism also appeared extreme when its adherents declared that only an end to slavery was acceptable, and that reforms were pro-slavery.)

Marx understood modern society as a state of "permanent revolution," in perpetual, innovating movement. Postmodernity brings more of the same, as accelerating change renders everything human (such as our closest relationships) frail and undone. The reality of this motion and fluidity has been raised to a virtue by postmodern thinkers, who celebrate undecidability as a universal condition. All is in flux, and context-free; every image or viewpoint is as ephemeral and as valid as any other.

This outlook is the postmodern totality, the position from which postmodernists condemn all other viewpoints. Postmodernism's historic ground is unknown to itself, because of a founding aversion to overviews and totalities. Unaware of Kaczynski's central idea (*Industrial Society and Its Future*, 1996) that meaning and freedom are progressively banished by modern technological society, postmodernists would be equally uninterested in the fact that Max Weber wrote the same thing almost a century before. Or that the movement of society, so described, is the historical truth of what postmodernists analyze so abstractly, as if it were a novelty they alone (partially) understand.

Shrinking from any grasp of the logic of the system as a whole, via a host of forbidden areas of thought, the anti-totality stance of these embarrassing frauds is ridiculed by a reality that is more totalized and global than ever. The surrender of the postmodernists is an exact reflection of feelings of helplessness that pervade the culture. Ethical indifference and aesthetic self-absorption join hands with moral paralysis, in the postmodern rejection of resistance. It is no surprise that a non-Westerner such as Ziauddin Sardan (*Postmodernism and the Other*, 1998) judges that postmodernism "preserves—indeed enhances—all the classical and modern structures of oppression and domination."

This prevailing fashion of culture may not enjoy much more of a shelf life. It is, after all, only the latest retail offering in the marketplace of representation. By its very nature, symbolic culture generates distance and mediation, supposedly inescapable burdens of the human condition. The self has always only been a trick of language, says Althusser. We are sentenced to be no more than the modes through which language autonomously passes, Derrida informs us.

The outcome of the imperialism of the symbolic is the sad commonplace that human embodiment plays no essential role in the functions of mind or reason. Conversely, it's vital to rule out the possibility that things have ever been different. Postmodernism resolutely bans the subject of origins, the notion that we were not always defined and reified by symbolic culture. Computer simulation is the latest advance in representation, its disembodied power fantasies exactly paralleling modernity's central essence.

The postmodernist stance refuses to admit stark reality, with discernible roots and essential dynamics. Benjamin's "storm" of progress is pressing forward on all fronts. Endless aesthetic-textual evasions amount to rank cowardice. Thomas Lamarre serves up a typical post-modern apologetic on the subject: "Modernity appears as a process or rupture and reinscription; alternative modernities entail an opening of otherness within Western modernity, in the very process of repeating or reinscribing it. It is as if modernity itself is deconstruction." (*Impacts of Modernities*, 2004).

Except that it isn't, as if anyone needed to point that out. Alas, deconstruction and detotalization have nothing in common. Deconstruction plays its role in keeping the whole system going, which is a real catastrophe, the actual, ongoing one.

The era of virtual communication coincides with the postmodern abdication, an age of enfeebled symbolic culture. Weakened and cheapened connectivity finds its analogue in the fetishization of ever-shifting, debased textual "meaning." Swallowed in an environment that is more and more one immense aggregate of symbols, deconstruction embraces this prison and declares it to be the only possible world. But the depreciation of the symbolic, including illiteracy and a cynicism about narrative in general, may lead in the direction of bringing the whole civilizational project into question. Civilization's failure at this most fundamental level is becoming as clear as its deadly and multiplying personal, social, and environmental effects.

"Sentences will be confined to museums if the emptiness of writing persists," predicted Georges Bataille. Language and the symbolic are the conditions for the possibility of knowledge, according to Derrida and the rest. Yet we see at the same time an ever-diminishing vista of understanding. The seeming paradox of an engulfing dimension of representation and a shrinking amount of meaning finally causes the former to become susceptible—first to doubt, then to subversion.

Husserl tried to establish an approach to meaning based on respecting experience/ phenomena just as it is delivered to us, before it is re-presented by the logic of symbolism. Small surprise that this effort has been a central target of postmodernists, who have understood the need to extirpate such a vision. Jean-Luc Nancy expresses this opposition succinctly, decreeing that "We have no idea, no memory, no presentiment of a world that holds man [sic] in its bosom" (*The Birth to Presence*, 1993). How desperately do those who collaborate with the reigning nightmare resist the fact that during the two million years before civilization, this earth was precisely a place that did not abandon us and did hold us to its bosom.

Beset with information sickness and time fever, our challenge is to explode the continuum of history, as Benjamin realized in his final and best thinking. Empty, homogenous, uniform time must give way to the singularity of the non-exchangeable present. Historical progress is made of time, which has steadily become a monstrous materiality, ruling and measuring life. The "time" of non-domestication, of non-time, will allow each moment to be full of awareness, feeling, wisdom, and re-enchantment. The true duration of things can be restored when time and the other mediations of the symbolic are put to flight. Derrida, sworn enemy of such a possibility, grounds his refusal of a rupture on the nature and allegedly eternal existence of symbolic culture: history *cannot* end, because the constant play of symbolic movement cannot end. This auto-da-fé is a pledge against presence, authenticity, and all that is direct, embodied, particular, unique, and free. To be trapped in the symbolic is only our current condition, not an eternal sentence.

It is language that speaks, in Heidegger's phrase. But was it always so? This world is over-full of images, simulations—a result of choices that may seem irreversible. A species has, in a few thousand years, destroyed community and created a ruin. A ruin called culture. The bonds of closeness to the earth and to each other—outside of domestication, cities, war, etc.—have been sundered, but can they not heal?

Under the sign of a unitary civilization, the possibly fatal onslaught against anything alive and distinctive has been fully unleashed for all to see. Globalization has in fact only intensified what was underway well before modernity. The tirelessly systematized colonization and uniformity, first set in motion by the decision to control and tame, now has enemies who see it for what it is and for the ending it will surely bring, unless it is defeated. The choice at the beginning of history was, as now, that of presence versus representation.

Gadamer describes medicine as, at base, the restoration of what belongs to nature. Healing as removing whatever works against life's wonderful capacity to renew itself. The spirit of anarchy, I believe, is similar. Remove what blocks our way and it's all there, waiting for us.

Globalization and its Apologists: AN ABOLITIONIST PERSPECTIVE

In its heyday in the American south, slavery never lacked for apologists. Writers, preachers, and planters chimed in to defend the peculiar institution as divinely ordained and justified by the racial superiority of whites over blacks. The Abolitionists, who burned the Constitution, hid fugitives, and attacked federal arsenals, were widely viewed as dangerous firebrands fit for prison or the gallows.

In hindsight, the word "slavery" connotes a world of oppression, violence, degradation, and resistance. The vile, deluded racism of slavery's 19th century apologists is unmistakable from our 21st century viewpoint, but how many see our century's version of slavery in a similarly revealing light?

In the name of progress, world development and empire are enslaving humankind and destroying nature, everywhere. The juggernaut known as globalization has absorbed nearly all opposition, overwhelming resistance by means of an implacable, universalizing system of capital and technology.[1] A sense of futility that approaches nihilism is now accepted as an inevitable response to modernity: "Whatever...." The poverty of theory is starkly illuminated in this fatalistic atmosphere. Academic bookshelves are loaded with tomes that counsel surrender and accommodation to new realities. Other enthusiasts have climbed onto the globalization bandwagon, or more commonly, were never not on board. From an abolitionist perspective, the response of most intellectuals to a growing planetary crisis consists of apologia in endless variations.

Patrick Brantlinger[2] suggests, for example, that in the "post-historical" age we have lost the ability to explain social change. But the reasons behind global change become evident to those who are willing to examine fundamental assumptions. The debasing of life in all spheres, now proceeding at a quickening pace, stems from the dynamics of civilization itself. Domestication of animals and plants, a process only 10,000 years old, has penetrated every square inch of the planet. The result is the elimination of individual and community autonomy and health, as well as the rampant, accelerating destruction of the natural world. Morris Berman, Jerry Mander, and other critics have described the "disenchantment" of a world subordinated to technological development. Civilization substitutes mediation for direct experience, distancing people from their natural surroundings and from each other. Ever greater anomie, dispersal, and loneliness pervade our lives. A parallel instrumentalism is at work in our ecosystems, transforming them into resources to be mined, and imperiling the entire biosphere.[3]

At base, globalization is nothing new. Division of labor, urbanization, conquest, dispossession, and diasporas have been part and parcel of the human condition since the beginning of civilization. Yet globalization takes the domesticating process to new levels. World capital now aims to exploit all available life; this is a defining and original trait of globalization. Early 20th century observers (Tönnies and Durkheim among them) noted the instability and fragmentation that necessarily accompanied modernization. These are only more evident in this current, quite possibly terminal stage. The project of integration through world control causes disintegration everywhere: more rootlessness, withdrawal, pointlessness. . . none of which have arrived overnight. The world system has become a high-tech imperialism. The new frontier is cyberspace. In the language of perennial empire, global powers issue their crusading, adventurous call to tame and colonize (or recolonize).[4]

Marshall McLuhan's "global village" concept is back in vogue, albeit with a clonal tinge to it, as everyone is designated to be part of a single global society. One interdependent McWorld, kept alive by the standardized sadness of a draining consumerism. It should be no surprise that among those who speak in the name of "anti-globaliza-tion" there are actually a growing number who in fact oppose it, whose perspective is that of de-globalization.

The "global village," subject to almost instantly worldwide epidem-ics[5] , has become a downright scary place. Since the 1980s the term "risk" has become pervasive in almost every discursive field or disci-pline in developed societies. The power of nation-states to "manage" risks has demonstrably declined, and individual anxiety has increased, with the spread of modernization and globalization.[6] This trajectory also brings growing disillusionment with representative government and a rising, if still largely inchoate anti-modern orientation. These outlooks have strongly informed anti-authoritarian movements in recent years. There is a perceived hollowness, if not malevolence, to basic social institutions across the board. As Manuel Castells puts it, "we can perceive around the world an extraordinary feeling of uneasi-ness with the current process of technology-led change that threatens to generate a widespread backlash."[7]

A technified world continues to proliferate, offering the promise of escape from the less and less attractive context of our lives. Hoping no-one realizes that technology is centrally responsible for impoverished reality, its hucksters spread countless enticements and promises, while it continues to metastasize. Net/Web culture (a revealing nomencla-ture) is a prime example, extending its deprived version of social exis-tence via virtual space. Now that embedded, face-to-face connectivity is being so resolutely annihilated, it's time for virtual community.

According to Rob Shields' chilling formulation, "the presence of absence is virtual."[8] "Community" is unlike any other in human mem-ory; no real people are present and no real communication takes place. In convenient, disembodied virtual community, one shuts people off at the click of a mouse to "go" elsewhere. Pseudo-community moves forward on the ruins of what is left of actual connections. Senses and sensuality diminish apace;[9] "responsibility" is interred in the expanding postmodern Lost Words Museum. Shriveled opposition and fatalistic, resigned shirkers forget that anti-slavery abolitionists, once a tiny minority, refused to quit and eventually prevailed.

Certainly none of this has happened overnight. The AT&T telephone commercial/exhortation of some years back, "Reach out and touch someone," offered human contact but concealed the truth that such technology has in fact been crucial in taking us ever further from that contact. Direct experience is replaced by mediation and simulation. Digitized information supplants the basis of actual closeness and possible trust among interacting physical beings. According to Boris Groys, "We just have to deal with the fact that we can no longer believe our eyes, our ears. Everyone who has worked with a computer knows that."[10]

Globalization is likewise scarcely new on the economic and political scene. In the *Communist Manifesto* Marx and Engels predicted the emergence of a world market, based on growing production and consumption patterns of their day. The Spanish empire, 300 years earlier, was the first global power network.

Marx contended that every technology releases opposing possibilities of emancipation and domination. But somehow the project of a humanized technology has proven groundless and result-free; only technified humanity has come to pass. Technology is the embodiment of the social order it accompanies, and in its planetary advance transfers the fundamental ethos and values behind that technology. It never exists in a vacuum and is never value-neutral. Some alleged critics of technology speak, for example, of advancing "to a higher level of integration between humanity and nature."[11] This "integration" cannot avoid echoing the integration that is basic to civilization and its globalization; namely, the cornerstone institutions that integrate all into themselves. Foremost among them is division of labor.

A state of growing passivity in everyday life is one of the most basic developments. Increasingly dependent—even infantilized—by a technological life-world, and under the ever-more complete effective control of specialized expertise, the fractionated subject is vitiated by division of labor. That most fundamental institution, which defines complexity and has driven domination forward *ab origino*. Source of all alienation, "the subdivision of labour is the assassination of a people."[12] Adam Smith in the 18th century has perhaps never been excelled in his eloquent portrait of its mutilating, deforming, immiserating nature.[13]

It was the prerequisite for domestication,[14] and continues to be the motor of the Megamachine, to use Lewis Mumford's term. Division of labor underlies the paradigmatic nature of modernity (technology) and its disastrous outcome.

Although the wind is shifting in some quarters, it's somewhat baffling that theory has seldom put into question this institution (or domestication, for that matter). The latent desire for wholeness, simplicity, and the immediate or direct has been overwhelmingly dismissed as futile and/or irrelevant. "The task we now face is not to reject or turn away from complexity but to learn to live with it creatively," advises Mark Taylor.[15] We must "resist any simple nostalgia," counsels Katherine Hayles, while granting that "nightmare" may well describe what's been showing up lately.[16]

In fact, even more confounding than lack of interest in the roots and motive force undergirding the present desolation is the fairly widespread embrace of the prospect of more of the same. How is it possible to imagine good outcomes from what is clearly generating the opposite, in every sphere of life? Instead of a hideously cyborgian program delivering emptiness and dehumanization on a huge scale, Hayles, for instance, finds in the posthuman an "exhilarating prospect" of "opening up new ways of thinking about what being human means," while high-tech "systems evolve toward an open future marked by contingency and unpredictability."[17]

What's happening is that a "what we have lost" sensibility is being overwhelmed by a "what have we got to lose/try anything" orientation. This shift testifies profoundly to the depth of loss and defeat that civilization/patriarchy/industrialism/modernity has engineered. The magnitude of the surrender of these intellectuals has nullified their capacity for analysis or vision. For example, "Increasingly the question is not whether we will become posthuman, for posthumanity is already here."[18]

Technology as an injunction to forget, as a solvent of meaning,[19] finds its cultural voice in postmodernism. Articulated in the context of transnationalism whereby globalization renders its totalizing nature glaringly evident, postmodernism pursues its refusal of "any notion of representable or essential totality."[20] Helplessness reigns; there are no foundational places left from which to think about or resist the juggernaut. As Scott Lash states, "We can no longer step outside of the global communications flows to find a solid fulcrum for critique."[21] His misnamed *Critique of Information* announces total abdication: "My argument in this book is that such critique is no longer possible. The global information order itself has, it seems to me, erased and swallowed up the possibility of a space of critical reflection."[22]

With no ground from which to make judgments, the very viability of criteria dissolves; the postmodern thus becomes prey to every manner of preposterous and abject pronouncement. I. Bluhdorn, for example, simply waves the little matter of environmental catastrophe away: "To the extent that we manage to get used to (naturalize) the non-availability of universally valid normative standards, the ecological problem. . . simply dissolves."[23] The cynical acceptance of every continuing horror, clothed in aesthetized irony and implicit apathy.

Downright bizarre is the incoherent celebration of the marriage of the postmodern and the technological, summed up in a title: *The Postmodern Adventure: Science, Technology, and Cultural Studies at the Third Millennium.*[24] According to authors Steven Best and Douglas Kellner, "The postmodern adventure is just beginning and alternative futures unfold all around us." To speak of defending the particular against universalizing tendencies is a postmodern commonplace, but this is mocked by the eager acceptance of the most universalizing force of all, the homogenization machine which is technology.

Andrew Feenberg discusses the all-pervasive presence of technology, arguing that when the Left joins in the celebration of technological advances, the ensuing consensus leaves little to disagree about.

A leftist himself, Feenberg concludes that "we cannot recover wh... reification has lost by regressing to pretechnological conditions, to some prior unity irrelevant to the contemporary world."[25] But such "relevance" is what is really at issue. To remain committed to the "contemporary world" is precisely the foundationless foundation of complicity. Postmodernity as the realization or completion of universal technology, globalization's underlying predicate.

When the basics are ruled off-limits to contestation, the resulting evasion can have no liberatory consequence. Infatuation with surface, the marginal, the partial, etc. is typical. Postmodernism billed itself as subversive and destabilizing, but delivered only aesthetically. Emblematic of a period of defeat, the image consumes the event and we consume the images. The tone throughout Derrida's work, for instance, seems never far from mourning. The abiding sadness of Blanchot is also to the point. The postmodern, according to Geoffery Hartman, "suggests a disenchantment that is final, or self-perpetuating."[26]

The subject, in the current ethos, is seen on the one hand as an unstable, fragmented collection of positions in discourse—even as a mere effect of power, or of language—and on the other hand as part of a positive, pluralist array of alternatives. By avoiding examination of the main lines of domination, however, postmodernists blind themselves to the actual, deforming characteristics of technology and consumerism. The forgetful self of technology, buffeted by the ever-shifting currents of commodified culture, is hard-pressed to form an enduring identity. There is, in fact, an increasing distance between dominant global forces and the endangered coherent individual.

The high-tech network of the world system is completing the transformation of classes into masses, the erosion of group solidarity and autonomy, and the isolation of the self. As Bamyeh points out, these are the preconditions of modern mass democracies, as well as the basic political features of global modernity itself.[27] Meanwhile, participation in this setup dwindles, as a massified, standardized techno-world makes a joke of the idea that any of it could be changed on its own terms. Elections, for instance, are widely understood to

neaningless rituals, technicized and commodified
lation.[28] Fulfillment and freedom are fast evapo-
predominant note of social theory seems to be com-
uncritical. The subject is merely a shifting intersection of global
networks; "the I is a moment of complexity," says Mark Taylor in
unconcerned summary.[29]

Along with health-threatening obesity (largely due to the rapid
spread of "fast food" and other processed foods), depression has become
an international scourge. Among various consequences of development,
depression testifies directly to the loss of deeply important ingredients
of human happiness. But as Lyotard has it, "despair is taken as a disorder
to control, never as the sign of an irremediable lack."[30] Already the fourth
leading cause of disability in the U.S., depression is projected to take
second place by 2020. Despite the general reactionary focus on genetics
and chemical palliatives, depression has much more to do with the growing isolation of individuals within developed society. The figures about
declining social and civic membership or affiliation are relevant; the rise
of autism, binge drinking, and illiteracy betoken depression's progres-
sas an even more profound phenomenon. "At the time of the so-called
triumph of the West, why do so many people feel so crappy, so lonely, so
abandoned?" asks philosopher Bruce Wilshire.[31]

It should no longer appear paradoxical that a deepening malaise
co-exists with the escalating importance of expertise in managing
everyday life. People distrust the institutions, and have lost confidence
in themselves. Elissa Gootman's "Job Description: Life of the Party"
discusses hired party "motivators," professionals who guarantee successful socializing.[32] On a more serious note, instrumental rationality
penetrates our lives at ever-younger ages. Kids as young as two are now
routinely medicated for depression and insomnia.[33]

An array of postmodernisms and fundamentalisms seems to have
displaced belief in the future. Marcuse wondered whether narcissism's
yearning for completeness and perfection might not contain the germ
of a different reality principle. Even whether, contra Hegel, reconciliation could only happen outside of historical time.[34]

Such "critics" as there be (Chomsky, Derrida, Ricoeur, Plumwood, for example) call for a global governance/planning apparatus—under which, it must be said, the individual would have even less of a voice. Anti-totality Derrida wants a "New International," apparently ignorant of the actual zero degree of "democracy" that obtains in the current political jurisdictions. Such superficiality, avoidance, and illusion surely constitutes acceptance of the ongoing devastation. Of course, if statist regulation could be an answer it would necessarily be totalitarian. And it would be partial at best, because it would never indict any of civilization's motive forces, such as division of labor or domestication.

What is clear to some of us is that a turn away from the virtual, global networks of power, unlimited media, and all the rest is a necessity. A break with this worsening world toward embeddedness, the face-to-face, non-domination of nature and each other.

Todd Gitlin, while rejecting such a refusal as mere "wishfulness," is helpful on the subject: "So consistent abolitionists have little choice but to be root-and-branch, scorch-and-burn primitivists, scornful of the rewards of a consumer society, committed to cutting the links in the invisible chain connecting modern production, consumption, and the technologies implicated in both. Only unabashed primitivists can create postindustrial wholeness."[34]

12

Overman and Unabomber

Born a hundred years apart, the lives of Friedrich Nietzsche and Theodore Kaczynski contain some important parallels. Both refused extremely promising academic careers: Nietzsche in philology, Kaczynski in mathematics. Each tried to make the most of a basically solitary existence. "Philosophy, as I have understood and lived it to this day, is a life voluntarily spent in ice and high mountains," said Nietzsche in *Ecce Homo*. For Kaczynski, the ice and high mountains were a more literal description, given his years in a cabin in the Montana Rockies.

Leslie Chamberlain (*Nietzsche in Turin*, London, 1996) summed up Nietzsche's experience as "Godless, jobless, wifeless and homeless." Kaczynski wandered less, but the characterization fits him very closely, too. Both were failures in relating to women, and uninterested in considering the condition of women in society. The two were both menaced at times by illness and impoverishment. Each was betrayed by his only sibling: Nietzsche by his sister Elizabeth, who tampered with his writings when he was helpless to prevent her; Kaczynski by his brother David, who fingered him for the FBI.

Nietzsche's central concept was the will to power. Kaczynski's big idea was the power process.

Both extolled strength and attacked pity: Nietzsche with his critique of Christianity as an unhealthy "slave morality," Kaczynski in terms of leftism as a dishonest projection of personal weakness.

Each developed, at base, a moral psychology, although Kaczynski is not limited to a psychology.

Nietzsche's analysis is contained within culture. His quest for a regeneration of the human spirit and the fulfillment of the individual is essentially aesthetic. Art, in many ways, replaced God for him. His post-Christian artistic vision is the measure of the Dionysian "revaluation of values." "What matters most...is always culture" (*Twilight of the Gods*).

There is no getting around Nietzsche's belief in hierarchy, his justification of rank and exploitation. Kaczynski's anarchist vision called for free community, decentralized to the point of face-to-face interaction.

Kaczynski, like Nietzsche, also desires virility over decadence, but saw that this can only be realized in terms of a social transformation. In *Beyond Good and Evil*, Nietzsche blamed "the democratization of Europe" for what he saw as a herd mentality. In *Industrial Society and Its Future*, Kaczynski recognized that a much deeper change than the political (not to mention the aesthetic) would be needed for the individual to be fre and fulfilled. He understood the logic of industrialized life to be the obstacle, and called for its destruction. For him, how everyday life is experienced was a far more important factor than abstract values or aesthetic expression. Nietzsche and Kaczynski thus see the values crisis quite differently. Especially in the persona of Zarathustra, Nietzsche calls for personal redemption through an act of the will. Kaczynski does not overlook the context of the individual, the forces that frustrate his/her life at a basic level.

Nietzsche focused on German culture, e.g. the case of Wagner. Kaczynski examined the movement and consequences of an increasingly artificial and estranging global industrial order.

Nietzsche affirmed the free spirit in books such as *Human, All Too Human*, *Daybreak*, and *The Gay Science*, only to question the existence of free will in other texts. Kaczynski showed that individual autonomy is problematic in modern society, and that this problem is a function of that society.

Both Nietzsche and Kaczynski are seen as nihilists by many. The prevailing postmodern ethos elevates Nietzsche and ignores Kaczynski—largely because Nietzsche does not challenge society and Kaczynski does.

For postmodernism, the self is just a product, an outcome, nothing more than a surface effect. Nietzsche actually originated this stance (now also known as "the death of the subject"), which can be found in many of his writings. Kaczynski expressed a determinate autonomy and showed that the individual has not been extinguished. One can lament the end of the sovereign individual and lapse into postmodern passivity and cynicism, or diagnose the individual's condition in society and challenge this condition, as Kaczynski did.

Freud's *Das Unbehagen in der Kultur*, translated in the 1920s as *Civilization and Its Discontents*, reads more literally as "what makes us uncomfortable about culture." Nietzsche never questioned culture itself. Kaczynski shed light on why industrialism, the ground of culture, must be overcome for health and freedom to exist.

13

Why Primitivism?

Debord biographer Anselm Giap[1] referred to the puzzle of the present, "where the results of human activity are so antagonistic to humanity itself," recalling a question posed nearly 50 years ago by Joseph Wood Krutch: "What has become of that opportunity to become more fully human that the 'control of nature' was to provide?"[2]

The general crisis is rapidly deepening in every sphere of life. On the biospheric level, this reality is so well-known that it could be termed banal, if it weren't so horrifying. Increasing rates of species extinctions, proliferating dead zones in the world's oceans, ozone holes, disappearing rainforests, global warming, the pervasive poisoning of air, water, and soil, to name a few realities.

A grisly link to the social world is widespread pharmaceutical contamination of watersheds.[3] In this case, destruction of the natural world is driven by massive alienation, masked by drugs. In the U.S., life-threatening obesity is sharply rising, and tens of millions suffer from serious depression and/or anxiety.[4] There are frequent eruptions of multiple homicides in homes, schools, and workplaces, while the suicide rate among young people has tripled in recent decades.[5] Fibromyalgia, chronic fatigue syndrome, and other "mystery" /psychosomatic illnesses have multiplied, vying with the emergence of new diseases with known physiological origins: Ebola, Lassa fever, AIDS, Legionnaires' disease. The illusion of technological mastery is mocked by the antibiotic-resistant return of TB and malaria, not to mention outbreaks of E coli, mad cow disease, West Nile virus, etc. Even a cursory survey of contemporary psychic immiseration would require many pages. Barely suppressed anger, a sense of emptiness, corrosion of belief in institutions across the board, high stress levels, all contribute to what Kornoouh has called "the growing fracture of the social bond."[6]

Today's reality keeps underlining the inadequacy of current theory and its overall retreat from any redemptive project. It seems undeniable that's what's left of life on earth is being taken from us. Where is the depth of analysis and vision to match the extremity of the human condition and the fragility of our planet's future? Are we simply only with a totalizing current of degradation and loss?

The crisis is diffuse, but at the same time it is starkly visible on every level. One comes to agree with Ulrich Beck that "people have begun to question modernity... its premises have begun to wobble. Many people are deeply upset over the house-of-cards character of superindustrialism."[7] Agnes Heller observed that our condition becomes less stable and more chaos-prone the further we move away from nature, contrary to the dominant ideology of progress and development.[8] With disenchantment comes a growing sense that something different is urgently needed.

For a new orientation the challenge is at a depth that theorists have almost entirely avoided. To go beyond the prospectless malaise, the collapse of social confidence so devastatingly expressed in *Les Particules Elémentaires* (Michel Houlebecq's end-of-the-millennium novel),[9] the analytical perspective simply must shift in a basic way. This consists, for openers, in refusing Foucault's conclusion that human capacities and relations are inescapably technologized.[10]

As Eric Vogelin put it, "The death of the spirit is the price of progress."[11] But if the progress of nihilism is identical to the nihilism of progress, whence comes the rupture, the caesura? How to pose a radical break from the totality of progress, technology, modernity?

A quick scan of recent academic fads shows precisely where such a perspective has not been found. Frederic Jameson's apt formulation introduces the subject for us: "Postmodernism is what you have when the modernization process is complete and nature is gone for good."[12]

Postmodernism is the mirror of an ethos of defeat and reaction, a failure of will and intellect that has accommodated to new extremities of estrangement and destructiveness.[13] For the postmodernists, almost nothing can be opposed. Reality, after all, is so messy, shifting, complex, indeterminate; and oppositions are, of course, just so many false binarisms. Vacuous jargon and endless side-stepping transcend passé dualisms. Daniel White, for example, prescribed "a postmodern-ecological rubric that steps past the traditional either-or of the Oppressor and Oppressed..."[14]

In the consumerist realm of freedom, "this complex node, where technologies are diffused, where technologies are chosen," according to Mike Michael,[15] who can say if anything is at all amiss? Iain Chambers is an eloquent voice of postmodern abjectness, wondering whether alienation is not simply an eternal given: "What if alienation is a terrestrial constraint destined to frustrate the 'progress' introjected in all teleologies?... Perhaps there is no separate, autonomous alternative to the capitalist structuring of the present-day world. Modernity, the westernization of the world, globalization, are the labels of an economic, political and cultural order that is seemingly installed for the foreseeable future."[16]

The fixation on surface (depth is an illusion; so are presence and immediacy), the ban on unifying narratives and inquiry into origins, indifference to method and evidence, emphasis on effects and novelty, all find their expression in postmodern culture at large. These attitudes and practices spread everywhere, along with the technology it embraces without reservation. At the same time, though, there are signs that these trivializing and derivative recipes for "thought" may be losing their appeal.[17] An antidote to postmodern surrender has been made available, largely through what is known as the anti-globalization movement.

Jean-François Lyotard, who once thought that technologized existence offered options, has begun to write about the sinister development of a neo-totalitarian, instrumentalist imprisonment. In earlier essays he pointed to a loss of affect as part of the postmodern condition. More recently he has attributed that loss to techno-scientific hegemony. Crippled individuals are only part of the picture, as Lyotard portrays social effects of what can only be called instrumental reason, in pathological ascendance. And contra Habermas, this domination by instrumental reason is in no way challenged by "communicative action."[18] Referring to global urban development, Lyotard stated, "We inhabit the megalopolis only to the extent that we declare it uninhabitable. Otherwise, we are just lodged there." Also, "with the megalopolis, what is called the West realizes and diffuses its nihilism. It is called development."[19]

In other words, there may be a way out of the postmodern cul-de-sac, at least for some. Those still contained by the Left have a much different legacy of failure to jettison—one that obviously transcends the "merely" cultural. Discredited and dying as an actual alternative, this perspective surely also needs to go.

Hardt and Negri's *Empire*[20] will serve as a classic artifact of leftism, a compendium of the worn-out and left-over. These self-described communist militants have no notion whatsoever of the enveloping crisis. Thus they continue to seek "alternatives within modernity." They locate the force behind their communist revolution in "the new productive practices and the concentration of productive labor on the plastic and fluid terrain of the new communicative, biological, and mechanical technologies."[21] The leftist analysis valiantly upholds the heart of productionist marxism, in the face of ever-advancing, standardizing, destructive technique. Small wonder Hardt and Negri fail to consider the pulverization of indigenous cultures and the natural world, or the steady worldwide movement toward complete dehumanization.

Claude Kornoouh considers monstrous "the idea that progress consists in the total control of the genetic stock of all living beings." For him, this would amount to an unfreedom "that even the bloodiest totalitarianism of the 20th century was not able to accomplish."[22] Hardt and Negri would not shrink from such control, since they do not question any of its premises, dynamics, or preconditions.

It is no small irony that the militants of *Empire* stand exposed for the incomprehension of the trajectory of modernity by one of their opposite number, Oswald Spengler. As nationalist and reactionary that

Spengler was, *The Decline of the West* is the great masterwork of world history, and his grasp of Western civilization's inner logic is uncanny in its prescience.

Especially relevant here are Spengler's judgments, so many decades ago, concerning technological development and its social, cultural, and environmental impacts. He saw that the dynamic, promethean ("Faustian") nature of global civilization becomes fully realized as self-destructive mass society and equally calamitous modern technology. The subjugation of nature leads ineluctably to its destruction, and to the destruction of civilization. "An artificial world is permeating and poisoning the natural. The Civilization itself has become a machine that does, or tries to do everything in mechanical terms."[23] Civilized man is a "petty creator against Nature." "...This revolutionary in the world of life...has become the slave of his creature. The Culture, the aggregate of artificial, personal, self-made life-forms, develops into a close-barred cage...."[24]

Whereas Marx viewed industrial civilization as both reason incarnate and a permanent achievement, Spengler saw it as ultimately incompatible with its physical environment, and therefore suicidally transitory. "Higher Man is a tragedy. With his graves he leaves behind the earth a battlefield and a wasteland. He has drawn plant and animal, the sea and mountain into his decline. He has painted the face of the world with blood, deformed and mutilated it."[25] Spengler understood that "the history of this technics is fast drawing to its inevitable close."[26]

Theodor Adorno seemed to concur with elements of Spengler's thinking: "What can oppose the decline of the west is not a resurrected culture but the utopia that is silently contained in the image of its decline."[27] Adorno and Horkheimer's *Dialectic of Enlightenment*[28] has a critique of civilization at its core, with its focal image of Odysseus forcibly repressing the Sirens' song of eros. *Dialectic*'s central thesis is that "the history of civilization is...the history of renunciation."[29] As Albrecht Wellmer summed it up, "*Dialectic of Enlightenment* is the theory of an irredeemably darkened modernity."[30] This perspective, now continually augmented by confirming data, tends to render irrelevant both sources of theory and the logic of progress. If there is no escape from a condition we can understand all too well, what more is there to say?

Herbert Marcuse tried to lay out an escape route in *Eros and Civilization*,[31] by attempting to uncouple civilization from modernity. To preserve the "gains" of modernity, the solution is a "non-repressive" civilization. Marcuse would dispense with "surplus repression," implying that repression itself is indispensable. Since modernity depends on production, itself a repressive institution, redefining work as free play can salvage both modernity and civilization. I find this an implausible, even desperate defense of civilization. Marcuse fails to refute Freud's view that civilization cannot be reformed.

Freud argued in *Civilization and Its Discontents* that non-repressive civilization is impossible, because the foundation of civilization is a forcible ban on instinctual freedom and eros. To introduce work and culture, the ban must be permanently imposed. Since this repression and its constant maintenance are essential to civilization, universal civilization brings universal neurosis.[32] Durkheim had already noted that as humankind "advances" with civilization and the division of labor, "the general happiness of society is decreasing."[33]

As a good bourgeois, Freud justified civilization on the grounds that work and culture are necessary and that civilization enables humans to survive on a hostile planet. "The principal task of civilization, its actual raison d'etre, is to defend us against nature." And further, "But how ungrateful, how short-sighted after all to strive for the abolition of civilization! What would then remain would be a state of nature, and that would be far harder to bear."[34]

Possibly civilization's most fundamental ideological underpinning is Hobbes' characterization of the pre-civilized state of nature as "nasty, brutish, and short." Freud subscribed to this view, of course, as did Adorno and Horkheimer.

Since the mid-1960s there has been a paradigm shift in how anthropologists understand prehistory, with profound implications

for theory. Based on a solid body of archaeological ar
research, mainstream anthropology has abandoned
hypothesis. Life before or outside civilization is now
specifically as social existence prior to domestication o ....als and
plants. Mounting evidence demonstrates that before the Neolithic
shift from a foraging or gatherer-hunter mode of existence to an agricultural lifeway, most people had ample free time, considerable gender
autonomy or equality, an ethos of egalitarianism and sharing, and no
organized violence.

A (misleadingly-named) "Man the Hunter" conference at the
University of Chicago in 1966 launched the reversal of the Hobbes-
ian view, which for centuries had provided ready justification for all
the repressive institutions of a complex, imperializing Western culture.
Supporting evidence for the new paradigm has come forth from archaeologists and anthropologists such as Marshall Sahlins, Richard B. Lee,
Adrienne Zihlman, and many others; these studies are widely available,
and now form the theoretical basis for everything from undergraduate
courses to field research.

Archaeologists continue to uncover examples of how our Paleolithic forbears led mainly peaceful, egalitarian, and healthy lives for
about two million years. The use of fire to cook tuberous vegetables as
early as 1.9 million years ago, and long distance sea travel 800,000
years ago, are two findings among many that testify to an intelligence
equal to our own.[36]

Genetic engineering and imminent human cloning are just the
most current manifestations of a dynamic of control and domination
of nature that humans set in motion 10,000 years ago, when our ancestors began to domesticate animals and plants. In the 400 generations
of human existence since then, all of natural life has been penetrated
and colonized at the deepest levels, paralleling the controls that have
been ever more thoroughly engineered at the social level. Now we can
see this trajectory for what it really is: a transformation that inevitably brought all-enveloping destruction, that was in no way necessary.
Significantly, the worldwide archaeological record demonstrates that

many human groups tried agriculture and/or pastoralism and later gave them up, falling back on more reliable foraging and hunting strategies. Others refused for generations to adopt the domestication practices of close neighbors.

It is here that a primitivist alternative has begun to emerge, in theory and in practice.[37] To the question of technology must be added that of civilization itself. Ever-growing documentation of human prehistory as a very long period of largely non-alienated human life stands in stark contrast to the increasingly stark failures of untenable modernity.

In the context of his discussion of the limitations of Habermas, Joel Whitebook wrote, "It may be that the scope of and depth of the social and ecological crisis are so great that nothing short of an epochal transformation of world views will be commensurate with them."[38] Since that time, Castoriadis concluded that a radical transformation will "have to launch an attack on the division of labor in its hitherto known forms."[39] Division of labor, slowly emerging through prehistory, was the foundation of domestication and continues to drive the technological imperative forward.

The challenge is to disprove George Grant's thesis that we live in "a world where only catastrophe can slow the unfolding of the potentialities of technique,"[40] and to actualize Claude Kornoouh's judgment that revolution can only be redefined against progress.[41]

14

Second-Best Life:
Real Virtuality

Reams of empirical studies and a century or two of social theory have noticed that modernity produces increasingly shallow and instrumental relationships. Where bonds of mutuality, based on face-to-face connection, once survived, we now tend to exist in a depthless, dematerialized technoculture. This is the trajectory of industrial mass society, not transcending itself through technology, but instead becoming ever more fully realized.

In this context, it is striking to note that the original usage of "virtual" was as the adjectival form of "virtue". Virtual reality is not only the creation of a narcissistic subculture; it represents a much wider loss of identity and reality. Its essential goal is the perfect intimacy of human and machine, the eradication of difference between in-person and computer-based interaction.

Second Life. Born Again. Both are escape routes from a gravely worsening reality. Both the high-tech and the fundamentalist options are passive responses to the actual situation now engulfing us. We are so physically and socially distant from one another, and encroaching virtuality drives us ever further apart. We can choose to "live" as free-floating surrogates in the new, untrashed Denial Land of VR, but only if we embrace what Zizek called "the ruthless technological drive which determines our lives."[1]

Cyberspace means collapsing nature into technology, in the words of Allucquere Rosanne Stone; she notes that we are losing our grounding as physical beings.[2] The key response in the arid techno-world is, of course, more technology. Drug technology, for the 70 million Americans with insomnia; for the sexually dysfunctional males now dependent on Viagra, Cialis, etc.; for the depressed and anxious who no longer dream or feel.

And as this regime works to further flatten and suppress direct experience, Virtual Reality, its latest triumph, comes in to fill the void. Second Life, There, and whatever brand is next offer dream worlds, to a world denuded of dreams. In our time, "virtual bereavement" and "online grieving" are touted as superior to being present to comfort those who mourn;[3] where tiny infants are subjected to videos; where "teledildonics" delivers simulated sex to distant subjects.

"Welcome to Second Life. We look forward to seeing you in-world", the website promo beckons. Immersive and interactive, VR provides the space so unlike the reality its customers reject. For a few dollars, anyone can exist there as an "avatar" who will never grow old, bored, or over-weight. Wade Roush of *Technology Review* declares Second Life a success insofar as it is "less lonely and less predictable" than the life we have now.[4] This inversion of reality is the consolation of the supernatural of many religions, and serves a similar substitutive function.

Reality is disappearing behind a screen, as the separation of mind from body and nature intensifies. The technical means are being per-fected fairly quickly, making good on the promises of the early 1990s. At that time VR, despite much ballyhoo,[5] could not really deliver the goods. Fifteen or so years later, the technology of Second Life (for example) engages many users with a strong sense of physical presence and other pseudo-sensory effects. Virtual reality is now the definitive expression of the postmodern condition, perhaps best typified by the fact that nothing wild exists there, only what serves human consumption.

Foucault described the shift of power in modernity from sover-eignty to discipline, and an enormously technologized daily life has accelerated this shift.[6] Contemporary life is thoroughly surveilled and policed, to an unprecedented degree. But the weight and density of tech mediation create an even more defining reality, and a more profound stage of control. When the nature of experience, on a primary level, is so deeply altered, we are seeing a fundamental shift—a shift being extended everywhere, at an accelerating pace.

Virtual reality best typifies this movement, its simulations and robotic fantasies a cutting-edge component of the steadily advancing, universalizing, standardizing global culture. Sadly pertinent is Philip Zai's judgement that VR is the "metaphysical maturity of civilization".[7] All that is tangible, sensual, and earth-based corrodes and shrinks within technologically mediated existence.

Of course, there are forms of resistance to this latest efflorescence of the false. But a luddite reaction always seems to pale before the magnitude of what it faces. There is a very long, sedimented history behind every newest technological move, an unbroken chain of contingency. The leap involved in grasping new technics is made easier by the gradual impoverishment of human desires and aptitudes caused by the earlier innovations. The promise is, always, that more technology will bring improvement—which more accurately means, more technology will make up for what was lost in the preceding "advances". The only way out is to break this chain, by refusing its imperative.

Heidegger assailed the "objectification of all beings...brought into the disposal of representation and production," pointing out that "nature appears everywhere as the object of technology", and concluding that "World becomes object".[8] He also understood how technology changes our relation to things, a phenomenon underlined by virtual reality. "Talk of a respect for things is more and more unintelligible in a world that is becoming ever more technical. They are simply vanishing...,: remarked Gadamer.[9] Virtuality is certainly that "vanishing".

There has been in fact a recent counter-attack in favor of respecting things as such, in favor of freeing them from an instrumental status, at least on the philosophical plane. Titles such as *Things* (2004) and *The Lure of the Object* (2005) speak to this.[10] Desire for the authentic experience of "thingness" (Heidegger's term) is a rebuke to the pathological condition known as modernity, a realization that "accepting the otherness of things is the condition for accepting otherness as such."[11]

Immersion in virtual reality is a particularly virulent strain of this pathology because of the degree of interactivity and self-representation involved. Never has the built environment depended so crucially on our participation, and never before has this participation been so potentially totalizing. With its appeal as, literally, a second life, a second world, it is The Matrix—one that we ourselves are to continually pay to reproduce. Heinz Pagels' description of the symbolic, in general, certainly applies to virtual reality: in denying "the immediacy of reality and in creating a substitute we have but spun another thread in the web of our grand illusion."[12] This use of cyberspace takes representation to new levels of self-enclosure and self-domestication.

Spengler's survey of Western civilization led him to conclude that "an artificial world is permeating and poisoning the natural. The civilization itself has become a machine that does, or tries to do, everything in mechanical fashion."[13] Second Life, Google Earth, etc., with their graphics cards and broadband connections are sophisticated and enticing escape hatches, but it's still the same basic machine orientation. And VR, as David Gelernter happily proclaimed, "is the sort of instrument that modern life demands."[14]

Born of military research and the entertainment industry, Virtual Reality depends on us for its projected role throughout society. Real virtuality will be the norm when it infects various spheres, but only with our active consent. Wittgenstein felt that "it is not absurd e.g. to believe that the age of science and technology is the beginning of the end for humanity."[15] Science and technology are the greatest triumphs of civilization, and the point is more grimly apparent than ever.

15

Breaking Point?

The rapidly mounting toll of modern life is worse than we could have imagined. A metamorphosis rushes onward, changing the texture of living, the whole feel of things. In the not-so-distant past this was still only a partial modification; now the Machine converges on us, penetrating more and more to the core of our lives, promising no escape from its logic.

The only stable continuity has been that of the body, and that has become vulnerable in unprecedented ways. We now inhabit a culture, according to Furedi[1], of high anxiety that borders on a state of outright panic. Postmodern discourse suppresses articulations of suffering, a facet of its accommodation to the inevitability of further, systematic desolation. The prominence of chronic degenerative diseases makes a chilling parallel with the permanent erosion of all that is healthy and life-affirming inside industrial culture. That is, maybe the disease can be slowed a bit in its progression, but no overall cure is imaginable in this context—which created the condition in the first place.

As much as we yearn for community, it is all but dead. McPherson, Smith-Lovin and Brashears tell us that 19 years ago, the typical American had three close friends; now the number is two. Their national study also reveals that over this period of time, the number of people without one friend or confidant has tripled.[2] Census figures show a correspondingly sharp rise in single-person households, as the techno-culture—with its vaunted "connectivity"—grows steadily more isolating, lonely and empty.

In Japan "people simply aren't having sex" and the suicide rate has been rising rapidly.[3] *Hikikimori*, or self-isolation, finds over a million young people staying in their rooms for years. Where the technoculture is most developed, levels of stress, depression and anxiety are highest.

Questions and ideas can only become currents in the world insofar as reality, external and internal, makes that possible. Our present state, devolving toward catastrophe, displays a reality in unmistakable terms. We are bound for a head-on collision between urgent new questions and a totality—global civilization—that can provide no answers. A world that offers no future, but shows no signs of admitting this fact, imperils its own future along with the life, health, and freedom of all beings on the planet. Civilization's rulers have always squandered whatever remote chances they had to prepare for the end of life as they know it, by choosing to ride the crest of domination, in all its forms.

It has become clear to some that the depth of the expanding crisis, which is as massively dehumanizing as it is ecocidal, stems from the cardinal institutions of civilization itself. The discredited promises of Enlightenment and modernity represent the pinnacle of the grave mistake known as civilization. There is no prospect that this Order will renounce that which has defined and maintained it, and apparently little likelihood that its various ideological supporters can face the facts. If civilization's collapse has already begun, a process now unofficially but widely assumed, there may be grounds for a widespread refusal or abandonment of the reigning totality. Indeed, its rigidity and denial may be setting the stage for a cultural shift on an unprecedented scale, which could unfold rapidly.

Of course, a paradigm shift away from this entrenched, but vulnerable and fatally flawed system is far from unavoidable. The other main possibility is that too many people, for the usual reasons (fear, inertia, manufactured incapacity, etc.) will passively accept reality as it is, until it's too late to do anything but try to deal with collapse. It's noteworthy that a growing awareness that things are going wrong, however inchoate and individualized, is fuelled by a deep, visceral unease and in many cases, acute suffering. This is where opportunity resides. From this new perspective that is certainly growing, we find the work of confronting what faces us as a species, and removing the barriers to planetary survival. The time has come for a wholesale indictment of civilization and mass society. It is at least possible that, in various modes, such a judgment can undo the death-machine before destruction and domestication inundate everything.

Although what's gone before helps us understand our current plight, we now live in obvious subjection, on a plainly greater scale than heretofore. The enveloping techno-world that is spreading so rapidly suggests movement toward even deeper control of every aspect of our lives. Adorno's assessment in the 1960s is proving valid today: "Eventually the system will reach a point—the word that provides the social cue is 'integration'—where the universal dependence of all moments on all other moments makes the talk of causality obsolete. It is idle to search for what might have been a cause within a monolithic society. Only that society itself remains the cause."[4]

A totality that absorbs every "alternative" and seems irreversible. Totalitarian. It is its own justification and ideology. Our refusal, our call to dismantle all this, is met with fewer and fewer countervailing protests or arguments. The bottom-line response is more along the lines of "Yes, your vision is good, true, valid; but this reality will never go away."

None of the supposed victories over inhumanity have made the world safer, not even just for our own species. All the revolutions have only tightened the hold of domination, by updating it. Despite the rise and fall of various political persuasions, it is always production that has won; technological systems never retreat, they only advance. We have been free or autonomous insofar as the Machine requires for its functioning.

Meanwhile, the usual idiotic judgments continue. "We should be free to use specific technologies as tools without adopting technology as lifestyle."[5] "The worlds created through digital technology are real to the extent that we choose to play their games."[6]

Along with the chokehold of power, and some lingering illusions about how modernity works, the Machine is faced with worsening prospects. It is a striking fact that those who manage the dominant organization of life no longer even attempt answers or positive projections. The most pressing "issues" (e.g. Global Warming) are simply ignored, and propaganda about Community (the market plus isolation), Freedom (total surveillance society), the American Dream (!) is so false that it cannot be expected to be taken seriously.

As Sahlins pointed out, the more complex societies become, the less they are able to cope with challenges. The central concern of any state is to preserve predictability; as this capacity visibly fails, so do that state's chances of survival. When the promise of security wanes, so does the last real support. Many studies have concluded that various ecosystems are more likely to suffer sudden catastrophic collapse, rather than undergo steady, predictable degradation. The mechanisms of rule just might be subject to a parallel development.

In earlier times there was room to maneuver. Civilization's forward movement was accompanied by a safety valve: the frontier. Large-scale expansion of the Holy Roman Empire eastward during the 12th–14th centuries, the invasion of the New World after 1500, the Westward movement in North America through the end of the 19th century. But the system becomes indebted to structures accumulated during these movements. We are hostages, and so is the whole hierarchical ensemble. The whole system is busy, always in flux; transactions take place at an ever-accelerating rate. We have reached the stage where the structure relies almost wholly on the co-optation of forces that are more or less outside its control. A prime example is the actual assistance given by modernizing leftist regimes in South America. The issue is not so much that of the outcome of neo-liberal economics in particular, but of the success of the left in power at furthering self-managed capital and co-opting indigenous resistance into its orbit, in the service of enforcing productivist logic in general.

But these tactics do not outweigh the fact of an overall inner rigidity that puts the future of techno-capital at grave risk. The name of the crisis is modernity itself, its contingent, cumulative weight. Any regime today is in a situation where every "solution" only deepens the engulfing problems. More technology and more coercive force are the only resources to fall back on. The "dark side" of progress stands revealed as the definitive face of modern times.

Theorists such as Giddens and Beck admit that the outer limits of modernity have been reached, so that disaster is now the latent characteristic of society. And yet they hold out hope, without predicating basic change, that all will be well. Beck, for instance, calls for a democratization of industrialism and technological change—carefully avoiding the question of why this has never happened.

There is no reconciliation, no happy ending within this totality, and it is transparently false to claim otherwise. History seems to have liquidated the possibility of redemption; its very course undoes what has been passing as critical thought. The lesson is to notice how much must change to establish a new and genuinely viable direction. There never was a moment of choosing; the field or ground of life shifts imperceptibly in a multitude of ways, without drama, but to vast effect. If the solution were sought in technology, that would of course only reinforce the rule of modern domination; this is a major part of the challenge that confronts us.

Modernity has reduced the scope allowed for ethical action, cutting off its potentially effective outlets. But reality, forcing itself upon us as the crisis mounts, is becoming proximal and insistent once again. Thinking gnaws away at everything, because this situation corrodes everything we have wanted. We realize that it is up to us. Even the likelihood of a collapse of the global techno-structure should not lure us away from acknowledgement of our decisive potential roles, our responsibility to stop the engine of destruction. Passivity, like a defeated attitude, will not bring forth deliverance.

We are all wounded, and paradoxically, this estrangement becomes the basis for communality. A gathering of the traumatized may be forming, a spiritual kinship demanding recovery. Because we can still feel acutely, our rulers can rest no more easily than we do. Our deep need for healing means that an overthrow must take place. That alone would constitute healing. Things "just go on", creating the catastrophe on every level. People are figuring it out:

that things just go on is, in fact, the catastrophe.

Melissa Holbrook Pierson expressed it this way: "Suddenly now it hits, bizarrely easy to grasp. We are inexorably heading for the Big Goodbye. It's official! The unthinkable is ready to be thought. It is finally in sight, after all of human history behind us. In the pit of what is left of your miserable soul you feel it coming, the definitive loss of home, bigger than the cause of one person's tears. Yours and mine, the private sob, will be joined by a mass crying...."[7]

Misery. Immiseration. Time to get back to where we have never quite given up wanting to be. "Stretched and stretched again to the elastic limit at which it will bear no more," in Spengler's phrase.

Enlightenment thought, along with the Industrial Revolution, began in late 18th century Europe, inaugurating modernity. We were promised freedom based on conscious control over our destiny. But Enlightenment claims have not been realized, and the whole project has turned out to be self-defeating. Foundational elements including reason, universal rights and the laws of science were consciously designed to jettison pre-scientific, mystical sorts of knowledge. Diverse, communally sustained lifeways were sacrificed in the name of a unitary and uniform, law-enforced pattern of living. Kant's emphasis on freedom through moral action is rooted in this context, along with the French encyclopedists' program to replace traditional crafts with more up-to-date technological systems. Kant, by the way, for whom property was sanctified by no less than his categorical imperative, favorably compared the modern university to an industrial machine and its products.

Various Enlightenment figures debated the pros and cons of emerging modern developments, and these few words obviously cannot do justice to the topic of Enlightenment. However, it may be fruitful to keep this important historical conjunction in mind: the nearly simultaneous births of modern progressive thought and mass production. Apt in this regard is the perspective of Min Lin: "Concealing the social origin of cognitive discourses and the idea of certainty is the inner requirement of modern Western ideology in order to justify or legitimate its position by universalizing its intellectual basis and creating a new sacred quasi-transcendence."[8]

Modernity is always trying to go beyond itself to a different state, lurching forward as if to recover the equilibrium lost so long ago. It is bent on changing the future—even its own—because it destroys the present. More modernity is needed to heal the wounds modernity inflicts!

With modernity's stress on freedom, modern enlightened institutions have in fact succeeded in nothing so much as conformity. Lyotard summed up the overall outcome: "A new barbarism, illiteracy and impoverishment of language, new poverty, merciless remodeling of opinion by media, immiseration of the mind, obsolescence of the soul."[9] Massified, standardizing modes, in every area of life, relentlessly re-enact the actual control program of modernity.

"Capitalism did not create our world; the machine did. Painstaking studies designed to prove the contrary have buried the obvious beneath tons of print."[10] Which is not in any way to deny the centrality of class rule, but to remind us that divided society began with division of labor. The divided self led directly to divided society. The division of labor is the labor of division. Understanding what characterizes modern life can never be far from the effort to understand technology's role in our everyday lives, just as it always has been. Lyotard judged that "technology wasn't invented by humans. Rather the other way around."[11]

Goethe's *Faust*, the first tragedy about industrial development, depicted its deepest horrors as stemming from honorable aims. The superhuman developer Faust partakes of a drive endemic to modernization, one which is threatened by any trace of otherness/ difference in its totalizing movement.

We function in an ever more homogeneous field, a ground always undergoing further uniformitization to promote a single, globalized techno-grid. Yet it is possible to avoid this conclusion by keeping one's focus on the surface, on what is permitted to exist on the margins. Thus some see Indymedia as a crucial triumph of decentralization, and free software as a radical demand. This attitude ignores the industrial basis of every high tech development and usage. All the "wondrous tools," including the ubiquitous and very toxic cell phone, are more related to eco-disastrous industrialization in China and India, for example, than to the clean, slick pages of *Wired* magazine. The salvationist claims of *Wired* are incredible in their disconnected, infantile fantasies. Its adherents can only maintain such gigantic delusions by means of deliberate blindness not only to technology's systematic destruction of nature, but to the global human cost involved: lives filled with toxicity, drudgery, and industrial accidents.

Now there are nascent protest phenomena against the all-encompassing universal system, such as "slow food," "slow cities," "slow roads". People would prefer that the juggernaut give pause and not devour the texture of life. But actual degradation is picking up speed, in its deworlding, disembedding course. Only a radical break will impede its trajectory. More missiles and more nukes in more countries is obviously another part of the general movement of the technological imperative. The specter of mass death is the crowning achievement, the condition of modernity, while the posthuman is the coming techno-condition of the subject. We are the vehicle of the Megamachine, not its beneficiary, held hostage to its every new leap forward. The technohuman condition looms, indeed. Nothing can change until the technological basis is changed, is erased.

Our condition is reinforced by those who insist—in classic postmodern fashion—that nature/culture is a false binarism. The natural world is evacuated, paved over, to the strains of the surrender-logic that nature has always been cultural, always available for subjugation. Koert van Mensvoort's "Exploring Next Nature" exposes the domination of nature logic, so popular in some quarters: "Our next nature will consist of what used to be cultural."[12] Bye-bye, non-engineered reality. After all, he blithely proclaims, nature changes with us.

This is the loss of the concept of nature altogether—and not just the concept! But the sign "nature" certainly enjoys popularity, as the substance is destroyed: "exotic" third world cultural products, natural ingredients in food, etc. Unfortunately, the nature of experience is linked to the experience of nature. When the latter is reduced to an insubstantial presence, the former is disfigured. Paul Berkett cites Marx and Engels to the effect that with communism people will "not only feel but also know their oneness with nature," that communism is "the unity of being of man with nature." Industrial-technological overcoming as its opposite—what blatant productionist rubbish. Leaving aside the communism orientation, however, how much of today's Left disagrees with the marxian ode to mass production? Where is any serious critique (one with consequences) of the massified, standardized Dead Zone that continues to spread everywhere?

A neglected insight in Freud's *Civilization and its Discontents* is the suggestion that a deep, unconscious "sense of guilt produced by civilization" causes a growing malaise and dissatisfaction.[14] Adorno saw that relevant to "the catastrophe that impends is the supposition of an irrational catastrophe in the beginning. Today the thwarted possibility of something other has shrunk to that of averting catastrophe in spite of everything."

The original, qualitative, utter failure for life on this planet was the setting in motion of civilization. Enlightenment—like the Axial Age world religions 2000 years before—supplied transcendence for the next level of domination, an indispensable support for industrial modernity. But where would one now find the source of a transcending, justifying framework for new levels of rapacious development? What new realm of ideas and values can be conjured up to validate the all-encompassing ruin of late modernity? There is none. Only the system's own inertia; no answers, and no future.

Meanwhile our context is that of a sociability of uncertainty. The moorings of day-to-day stability are being unfastened, as the system begins to show multiple weaknesses. When it can no longer guarantee security, its end is near.

Ours is an incomparable historical vantage point. We can easily grasp the story of this universal civilization's malignancy. This understanding may be a signal strength for enabling a paradigm shift, the one that could do away with civilization and free us from the habitual will to dominate. A daunting challenge, to say the least; but recall the child who was moved to speak out in the face of collective denial. The Emperor was wearing nothing; the spell was broken.

October 8, 2006

16

Finding Our Way Back Home

The candles are flickering. Not only has modernity failed; it has become a threat to the survival of life on our planet. Genuine hope withers as we face modernity's final stage, a totally technicized existence. Faith in progress is gone, and the self is now disintegrating and dispersing into cyberspace.

Czeslaw Milosz spoke of the prevailing "logic of precipitous decline, one so remarkable in its constancy as to be without historical analogy."[1] A growing number of books tell us all about modernity's enveloping crisis, only to provide "answers" that in no way depart from modernity's framework. Postmodernism has tried to use "absence" as its foundational idea, arguing against the possibility of unmediated existence, or of attaining another, qualitatively different state of being. Postmodern thinkers dare not imagine or acknowledge the more likely foundation of our misery: human alienation or "absence" as cause and culmination of mass society on a global and unitary scale.

Any return to life outside of this one must remain forever closed to us. Such is the almost unanimous judgment, even though this ban is the very condition of civilization's continued existence.

The void at the core of all this is "addressed" by (among other things) consumption. The insatiable hunger of modernity is built-in; no amount of re-shuffling the deck—by the Left, for example—can change this. Buying, working, anxiety, stress, depression are inherent, and exhibit an ever-deepening spiral. Consumption of the very life of the land is the way of civilization. Once people felt that historical development redeemed us from the meaninglessness of cycles of consumption. No longer. To consume is to devour, to hunger always in vain, and there is nothing redeeming about it.

But where all is integration into the totality, there is also a fear of totality and a different sort of hunger—a yearning for spiritual depth and renewal. Our sense of an overwhelming loss of wholeness, meaning, and authenticity drives a new impulse. The age of politics is over, because too many people know how pointless it is to continue choosing within the prevailing model for living. A few still assert that philosophy must chug along, running on conceptual resources compatible with our situation in a thoroughly disenchanted world. But more and more people know that this is not enough; that it is, in fact, intolerable.

Where do we look for rescue? Our predicament points us toward a solution. The crisis of modernity is, in a very basic sense, a failure of vision in which our disembodied life-world has lost its "place" in existence. We no longer see ourselves within the webs and cycles of nature. The loss of a direct relationship to the world terminates a once universal human understanding of our oneness with the natural world. The principle of relatedness is at the heart of indigenous wisdom: traditional intimacy with the world as the immanent basis of spirituality. This understanding is an essential and irreplaceable foundation of human health and meaningfulness.

Only if these ties are re-established can a spirituality that matters return. Religion, a contrived human projection (cf. Feuerbach, Nietzsche, Freud, et al.) is no substitute. Tom Porter, Mohawk, put it succinctly: "Now we have religion whereas before we had a way of life."[2] Every ideology is likewise founded on that loss of kinship with a prior world, that primary alienation from nature.

Novalis and Nietzsche both referred to philosophy as a kind of homesickness, the desire to be everywhere at home. Now we are nowhere at home. But our lament, our mourning for lost connection is only pointless suffering until it is linked to a reversal of our course. Modernity takes us ever further from home, and denies that any homecoming is conceivable. Yet the ensuing nihilism is urging into presence a new spiritual dimension that uncovers pathways that could lead us back—pathways that have been systematically hidden from sight during 10,000 years of civilization.

It is becoming too obvious that what bars our way is our failure to put an end to the reigning institutions and illusions. We must allow ourselves to see what has happened to us, including the origins of this disaster. At the same time we realize that true revolt is inspired by the realization that it is *not* impossible to bring the disaster to a halt, to imagine and strike out in new directions—to find our way back home.

Productionism or the primitive future, two materialities. One brought on by the extinguishing of spirit, the other by embracing spirit in its earth-based reality. The voluntary abandonment of the industrial mode of existence is not self-renunciation, but a healing return. Turning from this world's present state and direction, let's look for guidance from those who have continued to live spiritually within nature. Their example shows what we need to make our way to what still awaits, all around us.

Notes

I : TOO MARVELOUS FOR WORDS

[1] Paul Feyerabend, *Conquest of Abundance: A Tale of Abstraction versus the Richness of Being* (Chicago: University of Chicago Press, 1999), p. 270.

[2] Terence H. Hawkes, *Structuralism and Semiotics* (London: Methuen, 1977), pp. 149, 26.

[3] Michael Baxandall, *Giotto and the Orators* (Oxford: Clarendon Press, 1971), p. 44.

[4] Paul Feyerabend, *Killing Time* (Chicago: University of Chicago Press, 1995), p. 179.

[5] Susanne K. Langer, *Philosophy in a New Key* (Cambridge: Harvard University Press, 1942), p. 75.

[6] Ernest Jones, cited in Dan Sperber, *Rethinking Symbolism* (Cambridge: Cambridge University Press, 1975), p. 43.

[7] Edward Sapir, "The Emergence of the Concept of Personality in a Study of Cultures," *Journal of Social Psychology 5* (1934), pp 408-415.

[8] For example, Johann Gottfried Herder, *Treatise on the Origin of Language*.

[9] Michel Foucault, *The Archaeology of Knowledge*, translated by A.M.Sheridan Smith (New York: Pantheon, 1972), p. 216.

[10] Terrence W. Deacon, *The Symbolic Species* (New York: W.W. Norton, 1997), passim.

[11] Ernst Cassirer, *Language and Myth* (New York: Dover, 1953), pp 45-49.

[12] Sigmund Freud, *Moses and Monotheism, The Standard Edition of the Complete Works* (London: The Hogarth Press, 1964), p. 114.

[13] Marlene Nourbese Philip, *Looking for Livingstone* (Stratford, Ontario: Mercury Press, 1991), p. 11.

[14] Dan Sperber, "Anthropology and Psychology: Towards an Epidemiology of Representations," *Man 20* (1985), pp 73-89.

[15] The major rise in the incidence of autism is not metaphorical. Autism as a retreat from symbolic interaction seems to be a terrible commentary on its unfulfilling nature. It may not be coincidental that autism first appears in the medical literature in 1799, as the Industrial Revolution was taking off.

[16] Geert Lovink, *Uncanny Networks* (Cambridge: The MIT Press, 2002), p. 260.

[17] George Steiner, *Grammars of Creation* (New.Haven: Yale University Press, 2001), p. 3.

2: PATRIARCHY, CIVILIZATION AND THE ORIGINS OF GENDER

[1] Camille Paglia, *Sexual Personae: Art and Decadence from Nefertiti to Emily Dickinson* (Yale University Press: New Haven, 1990), p. 38.

[2] Ursula Le Guin, "Women/Wildness," in Judith Plant, ed., *Healing the Wounds* (New Society: Philadelphia, 1989), p. 45.

[3] Sherry B. Ortner, *Making Gender: the Politics and Erotics of Culture* (Beacon Press: Boston, 1996), p. 24. See also Cynthia Eller, *The Myth of Matriarchal Prehistory: Why an Invented Past Won't Give Women a Future* (Beacon Press: Boston, 2000).

[4] For example, Adrienne L. Zihlman and Nancy Tanner, "Gathering and Hominid Adaptation," in Lionel Tiger and Heather Fowler, eds., *Female Hierarchies* (Beresford: Chicago, 1978); Adrienne L. Zihlman, "Women in Evolution," *Signs 4* (1978); Frances Dahlberg, *Woman the Gatherer* (Yale University Press: New Haven, 1981); Elizabeth Fisher, *Woman's Creation: Sexual Evolution and the Shaping of Society* (Anchor/ Doubleday: Garden City NY, 1979).

[5] James Steele and Stephan Shennan, eds., *The Archaeology of Human Ancestry* (Routledge: New York, 1995), p. 349. Also, M. Kay Martin and Barbara Voorhies, *Female of the Species*

(Columbia University Press: New York, 1975), pp 210-211, for example.

⁶ Leacock is among the most insistent, claiming that whatever male domination exists in surviving societies of this kind is due to the effects of colonial domination. See Eleanor Burke Leacock, "Women's Status in Egalitarian Society," *Current Anthropology 19* (1978); and her *Myths of Male Dominance* (Monthly Review Press: New York, 1981). See also S. and G. Cafferty, "Powerful Women and the Myth of Male Dominance in Aztec Society," *Archaeology from Cambridge 7* (1988).

⁷ Joan Gero and Margaret W. Conkey, eds., *Engendering Archaeology* (Blackwell: Cambridge MA, 1991); C.F.M. Bird, "Woman the Toolmaker," in *Women in Archaeology* (Research School of Pacific and Asian Studies: Canberra, 1993).

⁸ Claude Meillasoux, *Maidens, Meal and Money* (Cambridge University Press: Cambridge, 1981), p. 16.

⁹ Rosalind Miles, *The Women's History of the World* (Michael Joseph: London, 1986), p. 16.

¹⁰ Zubeeda Banu Quraishy, "Gender Politics in the Socio-Economic Organization of Contemporary Foragers," in Ian Keen and Takako Yamada, eds., *Identity and Gender in Hunting and Gathering Societies* (National Museum of Ethnology: Osaka, 2000), p. 196.

¹¹ Jane Flax, "Political Philosophy and the Patriarchal Unconscious," in Sandra Harding and Merrill B. Hintikka, eds., *Discovering Reality* (Reidel: Dortrecht, 1983), pp 269-270.

¹² See Patricia Elliott, *From Mastery to Analysis: Theories of Gender in Psychoanalytic Feminism* (Cornell University Press: Ithaca, 1991), e.g. p. 105.

¹³ Alain Testart, "Aboriginal Social Inequality and Reciprocity," *Oceania 60* (1989), p. 5.

¹⁴ Salvatore Cucchiari, "The Gender Revolution and the Transition from Bisexual Horde to Patrilocal Band," in Sherry B. Ortner and Harriet Whitehead, eds., *Sexual Meanings: The Cultural Construction of Gender and Sexuality* (Cambridge University Press: Cambridge UK, 1984), p. 36. This essay is of great importance.

¹⁵ Olga Soffer, "Social Transformations at the Middle to Upper Paleolithic Transition," in Günter Brauer and Fred H. Smith, eds., *Replacement: Controversies in Homo Sapiens Evolution* (A.A. Balkema: Rotterdam 1992), p. 254.

¹⁶ Juliet Mitchell, *Women: The Longest Revolution* (Virago Press: London, 1984), p. 83.

¹⁷ Cucchiari, *op.cit.*, p. 62.

¹⁸ Robert Briffault, *The Mothers: the Matriarchal Theory of Social Origins* (Macmillan: New York, 1931), p. 159.

¹⁹ Theodore Lidz and Ruth Williams Lidz, *Oedipus in the Stone Age* (International Universities Press: Madison CT, 1988), p. 123.

²⁰ Elena G. Fedorova, "The Role of Women in Mansi Society," in Peter P. Schweitzer, Megan Biesele and Robert K. Hitchhock, eds., *Hunters and Gatherers in the Modern World* (Berghahn Books: New York, 2000), p. 396.

²¹ Steven Harrall, *Human Families* (Westview Press: Boulder CO, 1997), p. 89. "Examples of the link between ritual and inequality in forager societies are widespread," according to Stephan Shennan, "Social Inequality and the Transmission of Cultural Traditions in Forager Societies," in Steele and Shennan, *op.cit.*, p. 369.

²² Gayle Rubin, "The Traffic in Women," *Toward an Anthropology of Women* (Monthly Review Press: New York, 1979), p. 176.

²³ Meillasoux, *op.cit.*, pp 20-21.

²⁴ Cited by Indra Munshi, "Women and Forest: A Study of the Warlis of Western India,"

in Govind Kelkar, Dev Nathan and Pierre Walter, eds., *Gender Relations in Forest Societies in Asia: Patriarchy at Odds* (Sage: New Delhi, 2003), p. 268.

[25] Joel W. Martin, *Sacred Revolt: The Muskogees' Struggle for a New World* (Beacon Press: Boston, 1991), pp 99, 143.

[26] The production of maize, one of North America's contributions to domestication, "had a tremendous effect on women's work and women's health." Women's status "was definitely subordinate to that of males in most of the horticultural societies of [what is now] the eastern United States" by the time of first European contact. The reference is from Karen Olsen Bruhns and Karen E. Stothert, *Women in Ancient America* (University of Oklahoma Press: Norman, 1999), p. 88. Also, for example, Gilda A. Morelli, "Growing Up Female in a Farmer Community and a Forager Community," in Mary Ellen Mabeck, Alison Galloway and Adrienne Zihlman, eds., *The Evolving Female* (Princeton University Press: Princeton, 1997): "Young Efe [Zaire] forager children are growing up in a community where the relationship between men and women is far more egalitarian than is the relationship between farmer men and women" (p. 219). See also Catherine Panter-Brick and Tessa M. Pollard, "Work and Hormonal Variation in Subsistence and Industrial Contexts," in C. Panter-Brick and C.M. Worthman, eds., *Hormones, Health, and Behavior* (Cambridge University Press: Cambridge, 1999), in terms of how much more work is done, compared to men, by women who farm vs. those who forage.

[27] The Etoro people of Papua New Guinea have a very similar myth in which Nowali, known for her hunting prowess, bears responsibility for the Etoros' fall from a state of well-being. Raymond C. Kelly, *Constructing Inequality* (University of Michigan Press: Ann Arbor, 1993), p. 524.

[28] Jacques Cauvin, *The Birth of the Gods and the Origins of Nature* (Cambridge University Press: Cambridge, 2000), p. 133.

[29] Carol A. Stabile, *Feminism and the Technological Fix* (Manchester University Press: Manchester, 1994), p. 5.

[30] Carla Freeman, "Is Local:Global as Feminine:Masculine? Rethinking the Gender of Globalization," *Signs* 26 (2001).

3: ON THE ORIGINS OF WAR

[1] I Eibl-Eibesfelt, "Aggression in the !Ko-Bushmen," in Martin A. Nettleship, eds., *War, its Causes and Correlates* (The Hague: Mouton, 1975), p. 293.

[2] W.J. Perry, "The Golden Age," in *The Hibbert Journal XVI* (1917), p. 44.

[3] Arthur Ferrill, *The Origins of War from the Stone Age to Alexander the Great* (New York: Thames and Hudson, 1985), p. 16.

[4] Paul Taçon and Christopher Chippindale, "Australia's Ancient Warriors: Changing Depictions of Fighting in the Rock Art of Arnhem Land, N.T.," *Cambridge Archaeological Journal* 4:2 (1994), p. 211.

[5] Maurice R. Davie, *The Evolution of War: A Study of Its Role in Early Societies* (New Haven: Yale University Press, 1929), p. 247.

[6] A.L. Kroeber, *Handbook of the Indians of California: Bulletin 78* (Washington, D.C.: Bureau of American Ethnology, 1923), p. 152.

[7] Christopher Chase-Dunn and Kelly M. Man, *The Wintu and their Neighbors* (Tucson: University of Arizona Press, 1998), p. 101.

[8] Harry Holbert Turney-High, *Primitive War: Its Practice and Concepts* (Columbia: Univer-

sity of South Carolina Press, 1949), p. 229.

[9] Lorna Marshall, "Kung! Bushman Bands," in Ronald Cohen and John Middleton, eds., *Comparative Political Systems* (Garden City: Natural History Press, 1967), p. 17.

[10] George Bird Grinnell, "Coup and Scalp among the Plains Indians," *American Anthropologist* 12 (1910), pp. 296-310. John Stands in Timber and Margot Liberty make the same point in their *Cheyenne Memories* (New Haven: Yale University Press, 1967), pp. 61-69. Also, Turney-High, *op. cit.*, pp. 147, 186.

[11] Ronald R. Glassman, *Democracy and Despotism in Primitive Societies, Volume One* (Millwood, New York: Associated Faculty Press, 1986), p. 111.

[12] Emma Blake, "The Material Expression of Cult, Ritual, and Feasting," in Emma Blake and A. Bernard Knapp, eds., *The Archaeology of Mediterranean Prehistory* (New York: Blackwell, 2005), p. 109.

[13] Bruce M. Knauft, "Culture and Cooperation in Human Evolution," in Leslie Sponsel and Thomas Gregor, eds., *The Anthropology of Peace and Nonviolence* (Boulder: L. Rienner, 1994), p. 45.

[14] Roy A. Rappaport, *Pigs for the Ancestors: Ritual in the Ecology of a New Guinea People* (New Haven: Yale University Press, 1967), pp. 236-237.

[15] René Girard, *Violence and the Sacred*, translated by Patrick Gregory (Baltimore: Johns Hopkins University Press, 1977). Like Ardrey and Lorenz, Girard starts from the absurd view that all social life is steeped in violence.

[16] G. Lienhardt, *Divinity and Experience: The Religion of the Dinka* (Oxford: Oxford University Press, 1961), p. 281.

[17] Elizabeth Arkush and Charles Stanish, "Interpreting Conflict in the Ancient Andes: Implications for the Archaeology of Warfare," *Current Anthropology* 46:1 (February 2005), p. 16.

[18] *Ibid.*, p. 14.

[19] James L. Haley, *Apaches: A History and Culture Portrait* (Garden City, NY: Doubleday, 1981), pp. 95-96.

[20] Rappaport, *op.cit*, p. 234, for example.

[21] Quoted by Robert Kuhlken, "Warfare and Intensive Agriculture in Fiji," in Chris Gosden and Jon Hather, eds., *The Prehistory of Food: Appetites for Change* (New York: Routledge, 1999), p. 271. Works such as Lawrence H. Keeley, *War Before Civilization* (New York: Oxford University Press, 1996) and Pierre Clastres, *Archaeology of Violence* (New York: Semiotext(e), 1994), and Jean Guilaine and Jean Zammit, *The Origins of War: Violence in Prehistory* (Malden, MA: Blackwell, 2005) somehow manage to overlook this point.

[22] Verrier Elwin, *The Religion of an Indian Tribe* (London: Oxford University Press, 1955, p. 300.

[23] Jonathan Z. Smith, "The Domestication of Sacrifice," in Robert G. Hamerton-Kelly, ed., *Violent Origins* (Stanford: Stanford University Press, 1987), pp. 197, 202.

[24] Christine A. Hastorf and Sissel Johannessen, "Becoming Corn-Eaters in Prehistoric America," in Johannessen and Hastorf, eds., *Corn and Culture in the Prehistoric New World* (Boulder: Westview Press, 1994), especially pp. 428-433.

[25] Charles Di Peso, *The Upper Pima of San Cayetano de Tumacacori* (Dragoon, AZ: Amerind Foundation, 1956), pp. 19, 104, 252, 260.

[26] Christy G. Turner II and Jacqueline A. Turner, *Man Corn: Cannibalism and Violence in the Prehistoric American Southwest* (Salt Lake City: University of Utah Press, 1999), pp. 3, 460, 484.

[27] A.L. Kroeber, *Cultural and Natural Areas of Native North America* (Berkeley: University of California Press, 1963), p. 224.

[28] Harold B. Barclay, *The Role of the Horse in Man's Culture* (London: J.A. Allen, 1980), e.g. p. 23.

[29] Richard W. Howell, "War Without Conflict," in Nettleship, *op.cit.*, pp. 683-684.

[30] Betty J. Meggers, *Amazonia: Man and Culture in Counterfeit Paradise* (Chicago: Aldine Atherton, 1971), pp. 108, 158.

[31] Pierre Lemmonier, "Pigs as Ordinary Wealth," in Pierre Lemonnier, ed., *Technological Choices: Transformation in Material Cultures since the Neolithic* (London: Routledge, 1993), p. 132.

[32] Knauft, *op.cit.*, p. 50. Marvin Harris, *Cannibals and Kings* (New York: Random House, 1977), p. 39.

[33] Maurice Bloch, *Prey into Hunter: The Politics of Religious Experience* (Cambridge: Cambridge University Press, 1992), p. 88.

[34] The "rank-and-file" of organized labor is another product of these originals.

[35] Robert L. Carneiro, "War and Peace," in S.P. Reyna and R.E. Downs, eds., *Studying War: Anthropological Perspectives* (Langhorn, PA: Gordon and Breach, 1994), p. 12.

[36] Cited and discussed in Marshall Sahlins, *Stone Age Economics* (Chicago: Aldine, 19720, pp. 174, 182.

4: THE IRON GRIP OF CIVILIZATION: THE AXIAL AGE

[1] Jacques Cauvin, *The Birth of the Gods and the Origins of Agriculture* (Cambridge: Cambridge University Press, 2000), p.2.

[2] Karl Jaspers, *The Origin and Goal of History* (New Haven: Yale University Press, 1953), especially the first 25 pages.

[3] Christianity and Islam may be properly considered later spin-offs of this Axial period, their own natures already established some centuries earlier.

[4] Andrew Bosworth, "World Cities and World Economic Cycles," in *Civilizations and World Systems*, ed. Stephan K. Sanderson (Walnut Creek, CA: AltaMira Press, 1995), p. 214.

[5] Karl Jaspers, *Way to Wisdom* (New Haven: Yale University Press, 2003 [1951]), pp 98-99.

[6] Henry Bamford Parkes, *Gods and Men: The Origins of Western Culture* (New York: Vintage Books, 1965), p. 77.

[7] John Plott, *Global History of Philosophy*, vol. I (Delhi: Motilal Manarsidass, 1963), p. 8.

[8] Oswald Spengler, *The Decline of the West*, vol. II (New York: Alfred A. Knopf, 1928), p. 309

[9] Mircea Eliade, "Structures and Changes in the History of Religions," in *City Invincible*, eds. Carl H. Kraeling and Robert M. Adams (Chicago: University of Chicago Press, 1958), p. 365.

[10] *Ibid.*, pp 365-366. Karl Barth's leap into "the upper story of faith" has a similar sense; quoted in Seyyed Hossein Nasr, *Knowledge and the Sacred* (Albany: State University of New York, 1989), p. 48.

[11] Scott Atran, *In Gods We Trust: the Evolutionary Landscape of Religion* (New York: Oxford University Press, 2002), p. 57.

[12] S.N. Eisenstadt, "The Axial Age Breakthroughs," *Daedalus* 104 (1975), p. 13. "May the gods destroy that man who first discovered hours and who first set up a sundial here."—

Plautus, 3rd century B.C. Eisenstadt's is the best essay on the overall topic that I have found.

[13] The fate of domestic hand-loom weavers almost three millennia later comes to mind; the independent weaver household was overwhelmed by the factory system of the Industrial Revolution.

[14] It is a striking irony that Nietzsche named his archetypal "beyond good and evil" figure Zarathustra.

[15] Vilho Harle, *Ideas of Social Order in the Ancient World* (Westport, CT: Greenwood Press, 1998), p. 18.

[16] Spengler, *op. cit.*, pp 168, 205.

[17] V. Nikiprowetzky, "Ethical Monotheism," *Daedalus* 104 (1975), pp 80-81.

[18] Jacob Neusner, *The Social Studies of Judaism: Essays and Reflections, vol. 1* (Atlanta: Scholars Press, 1985), p. 71.

[19] Paolo Sacchi, *The History of the Second Temple Period* (Sheffield: Sheffield Academic Press Ltd., 2000), p. 87.

[20] *Ibid.*, pp 99-100.

[21] Frederick Klemm, *A History of Western Technology* (New York: Charles Scribners Sons, 1959), p. 28.

[22] Charles Singer, E. J. Holmyard and A.R. Hall, eds., *A History of Technology, vol. I* (Oxford: Clarendon Press, 1954), p. 408.

[23] C. Osborne Ward, *The Ancient Lowly, vol. I* (Chicago: Charles Kerr, 1888), Chapter V.

[24] Ludwig Edelstein, *The Idea of Progress in Classical Antiquity* (Baltimore: Johns Hopkins University Press, 1967), pp 15-16.

[25] *Ibid.*, p. 3.

[26] Romila Thapar, "Ethics, Religion, and Social Protest in India," *Daedalus* (104), 1975, p. 122. See also pp 118-121.

[27] For example, Vibha Tripathi, ed., *Archaeometallurgy in India* (Delhi: Sharada Publishing House, 1998), especially Vijay Kumar, "Social Implications of Technology."

[28] See Greg Bailey and Ian Mabbet, *The Sociology of Early Buddhism* (Cambridge: Cambridge University Press, 2004), pp 18-21. Bailey and Mabbet, it should be said, see more of the picture than just this aspect.

[29] Thapar, *op. cit.*, p. 125.

[30] Bailey and Mabbet, *op. cit.*, p. 3.

[31] Joseph Needham, *Science and Civilization in China, vol. 2* (Cambridge: Cambridge University Press, 1962), pp 99-100, 119.

[32] Spengler, *op. cit.*, p. 356.

5: ALONE TOGETHER: THE CITY AND ITS INMATES

[1] Joseph Grange, *The City: An Urban Cosmology* (Albany: State University of New York Press, 1999), p. xv.

[2] Edward Relph, *Place and Placelessness* (London: Pion Ltd., 1976), p. 6.

[3] Meanwhile, phenomena such as "Old Town" areas and historical districts distract from tedium and standardization, but also underline these defining urban characteristics. The patented superficiality of postmodern architecture underlines it as well.

[4] Max Weber, *The City*, translated by Don Martindale and Gertrud Neuwirth (Glencoe, IL: The Free Press, 1958), p. 11.

[5] *ibid.*, p. 21

[6] Lewis Mumford, *The Culture of Cities* (New York: Harcourt, Brace and Company, 1938), p. 6. For all of the valid historical content, Mumford can also lapse into absurdity, e.g. "the city should be an organ of love...." in *The City in History* (New York, Harcourt, Brace, 1961), p. 575.

[7] Michel de Certeau, *The Certeau Reader*, edited by Graham Ward (London: Blackwell Publishers, 2000), p. 103.

[8] Stanley Diamond, *In Search of the Primitive* (New Brunswick, NJ: Transaction Books, 1974), p. 7.

[9] *ibid.*, p. 1.

[10] Andrew Sherratt, *Economy and Society in Prehistoric Europe* (Princeton: Princeton University Press, 1997), p. 362.

[11] Arnold Toynbee, *Cities on the Move* (New York: Oxford University Press, 1970), p. 173.

[12] Nicolas Chamfort, quoted in James A. Clapp, *The City, A Dictionary of Quotable Thought on Cities and Urban Life* (New Brunswick, NJ: Center for Urban Policy Research, 1984), p. 51.

[13] Jean-Jacques Rousseau, *Emile*, translated by Allan Bloom (New York: Basic Books, 1979), p. 355.

[14] Richard Sennett, *Flesh and Stone: the Body and the City in Western Civilization* (New York: W.W. Norton, 1994), p. 23.

[15] Friedrich Engels, *The Condition of the Working Class in England* (St. Albans: Panther Press, 1969), p. 75.

[16] Alexis de Tocqueville, *Democracy in America* v. 2 (New York, Vintage, 1963), p. 141.

[17] Walter Benjamin, *Illuminations*, translated by Harry Zahn (New York: Schocken Books, 1969), p. 174.

[18] Kurt H. Wolff, *The Sociology of Georg Simmel* (New York: The Free Press, 1950), p. 413.

[19] Karl Marx, *Grundrisse* (New York, Vintage, 1973), p. 479.

[20] A typical and apposite work is Richard Harris, *Creeping Conformity: How Canada Became Suburban, 1900-1960* (Toronto: University of Toronto Press, 2004).

[21] Very pertinent is Michael Bull, *Sounding Out the City: Personal Stereos and the Management of Everyday Life* (New York, Oxford University Press, 2000).

[22] This is not only true in the West. In Arabic civilization, for example, madaniyya, or civilization, comes from madine, which means city.

[23] Julia Kristeva, *Strangers to Ourselves* (New York: Columbia University Press, 1991), p. 192.

[24] Toynbee, *op.cit.*, p. 196

[25] Jacques Ellul, *The Political Illusion* (New York: Alfred A. Knopf, 1967), p. 43.

[26] James Baldwin, *Nobody Knows My Name* (New York, The Dial Press, 1961), p. 65.

[27] Peter Marcuse and Ronald van Kempen, editors, *Of States and Cities: the Partioning of Urban Space* (New York, Oxford University Press, 2002), p. vii.

[28] John Habberton, *Our Country's Future* (Philadelphia: International Publishing Company, 1889), cited in Clapp, *op.cit.*, p. 105.

[29] Kai N. Lee, "Urban Sustainability and the Limits of Classical Environmentalism," in *Environment and Urbanization* 18:1 (April 2006), p. 9.

8: TWILIGHT OF THE MACHINES
SELECT BIBLIOGRAPHY

W.H. Auden, *The Double Man*, Random House, 1941.

Seyla Benhabib, "Feminism and the Question of Postmodernism," *The New Social Theory Reader*, ed. Steven Seidman and Jeffrey C. Alexander, Routledge, 2001

Marshall Berman, *The Twilight of American Culture*, W.W. Norton, 2000

Joseph Califano, "Group Calls Underage Drinking an Epidemic," *New York Times*, February 27, 2002

Jacques Ellul, *The Technological Society*, Vintage Books, 1964

Barbara Epstein, "Anarchism and the Anti-Globalization Movement," *Monthly Review*, September 2001

Dario Fo, "Italy, Mussolini's Ghost in these Times," *A-Infos web site*, March 3, 2002

Sigmund Freud, *Civilization and Its Discontents and The Future of an Illusion*, Hogarth Press, 1953

Chellis Glendinning, *My Name is Chellis and I'm in Recovery from Western Civilization*, Shambala Publications, 1994

David Graeber, "The New Anarchists," *New Left Review*, January-February 2002

Donna Haraway, *How Like a Leaf: Interview with T.N. Goodeve*, Routledge, 2000; *Modest Witness @Second Millennium*, Routledge, 1997; *Simians, Cyborgs and Women*, Routledge, 1991

Martin Heidegger, "The Question Concerning Technology," *Basic Writings*, ed. David Farrell Krell, Harper & Row, 1977

Esther Kaplan, "Keepers of the Flame," *Village Voice*, February 5, 2002

Henry Kissinger, *Does America Need a Foreign Policy?*, Simon & Schuster, 2001

Richard B. Lee, *Politics and History in Band Societies*, Cambridge University Press, 1982

Karl Marx, *The Grundrisse*, ed. David McLellan, Harper & Row, 1970

National Intelligence Council, *Global Trends 2015*, Langley, VA, 2000

Fredy Perlman, *Against His-story, Against Leviathan*, Black & Red, 1983

Marshall Sahlins, *Stone Age Economics*, Aldine, 1972

Scheffer, Carpenter, Foley, Folke, and Walker, "Catastrophic Shifts in Ecosystems," *Nature*, October 11, 2001

Oswald Spengler, *The Decline of the West*, Knopf, 1932

Joseph Tainter, *The Collapse of Complex Societies*, Cambridge University Press, 1988

Paul Virilio, *The Information Bomb*, Verso, 2000

Daniel R. White, *Postmodern Ecology*, State University of New York Press, 1998

John Zerzan, *Future Primitive*, Autonomedia and C.A.L. Press, 1994; Running On Emptiness, Feral House, 2002

Adrienne Zihlman, "Women as Shapers of the Human Adaptation," *Woman the Gatherer*, ed. F. Dahlberg, Yale University Press, 1981

9: EXILED FROM PRESENCE

[1] Alan Lightman, *Reunion* (New York: Pantheon Books, 2002).

[2] Stephen A. Erickson, "Absence, Presence and Philosophy," in *Phenomenology and Beyond: the Self and its Language*, eds. Harold A. Durfee and David F.T. Rodier (Dordrecht: Kluver Academic Publishers, 1989), p. 71.

[3] For example, *Organization for Economic Co-operation and Development, Emerging Risks in the 21st Century* (Paris: OECD, 2003), p. 3: "Large-scale disasters of the past few years—such as the terrorist attack of September 11, 2001, the appearance of previously unknown infectious diseases, unusually extensive flooding in large parts of Europe, devastating bushfires in Australia and violent ice storms in Canada—have brought home to OECD governments the realization that something new is happening.... Preparing to deal effectively with the hugely complex threats of the 21st century is a major challenge for decision makers in government and the private sector alike, and one that needs to be addressed as a matter of urgency." Even the ruling elites begin to see the magnitude of the crisis.

[4] Paul Shepard, *Man in the Landscape* (New York: Knopf, 1967), p. xxviii.

[5] Manuel de Landa, "Virtual Environments and the Emergence of Synthetic Reason," in *Flame Wars: the Discourse of Cyberculture*, ed. Mark Dery (Durham: Duke University Press, 1994, p. 284.

[6] It is certainly arguable that phenomenology as a method failed to "return to things themselves."

[7] Jacques Derrida, *Speech and Phenomena and Other Essays on Husserl's Theory of Signs* (Evanston: Northwestern University Press, 1973), p. 104.

[8] Jacques Derrida, "Sending: On Representation," *Social Research* 49:2 (Summer, 1982), p. 326.

[9] Quoted by Mikhail Harbameier, "Inventions of Writing," in *State and Society*, eds. John Glenhill, Barbara Bender and Mogens Trolle Larsen (London: Unwin Human, 1988), p. 265.

[10] Jacques Derrida, *Specters of Marx* (New York: Routledge, 1994).

[11] Paul Piccone, "Introduction," *TELOS* 124 (Summer 2002), p. 3.

[12] Gregory Ullmer, *Applied Grammatology* (Baltimore: Johns Hopkins University Press, 1995), pp 4, 5, 7, 10.

[13] George Landow, *Hypertext: The Convergence of Contemporary Critical Theory and Technology* (Baltimore: Johns Hopkins University Press, 1997).

[14] Martin Heidegger, "The Question Concerning Technology," *Basic Writings* (San Franciso: Harper, 1992), p. 33.

[15] Ellen Mortensen, *Teaching Thought: Ontology and Sexual Difference* (Boulder: Lexington Books, 2002), p. 117

[16] Timothy Lenoir, "The Virtual Surgeon: Operating on the Data in an Age of Medialization," in *Semiotic Flesh*, ed. Phillip Thurtle and Robert Mitchell (Seattle: University of Washington Press, 2002), p. 45. A similar grotesquerie, among many, is Andy Clark, *Natural-Born Cyborgs* (New York: Oxford University Press, 2003).

[17] Kathleen Woodward, "Distributed Systems: Of Cognition, Of the Emotions," in *Semiotic Flesh*, op.cit., p. 71.

[18] Jacquère Rosanne Stone, *The War of Desire and Technology at the Close of the Mecahnical Age* (Cambridge, MA: MIT Press, 1996), p. 183.

[19] Maggie Mort, Carl R. May, Tracy Williams, "Remote Doctors and Absent Patients: Acting at a Distance in Telemedicine?", *Science, Technology, & Human Values* 28:2 (Spring 2003),

pp 274-295.

[20] Theodor W. Adorno, *Negative Dialectics* (New York: Continuum, 1997), p. 19.

[21] Claude Merleau-Ponty, quoted by Leonard Lawlor, "Merleau-Ponty and Derrida: La Différance," in *Ecart and Différance: Merleau-Ponty and Derrida on Seeing and Writing* (Atlantic Highland, N.J.: Humanities Press, 1997), p. 106

II: GLOBALIZATION AND ITS APOLOGISTS: AN ABOLITIONIST PERSPECTIVE

[1] Anthony King, "Baudrillard's Nihilism and the End of Theory," *TELOS* 112 (Summer 1998).

[2] Patrick Brantlinger, "Apocalypse 2001; or, What Happens after Posthistory?" *Cultural Critique* 39 (Spring 1998).

[3] To speak in terms of a supposedly "unfinished project" of idealized modernity is bizarrely out of touch with reality.

[4] The globalization of the dominant culture is revealed in "The Culture of Globalization" by Klaus Müller (*Museum News*, May-June 2003). Eighteen of the world's leading museums, including the Louvre, the Metropolitan Museum of Art, and the Hermitage, announced in December 2002 that artifacts of various cultures must be available to an international public, and therefore would not be returned, even if they had been seized during colonial rule.

[5] Christine McMurran and Roy Smith, *Diseases of Globalization* (Earthscan Publications Ltd.: London, 2001) discusses deteriorating conditions.

[6] See Joost Van Loon, *Risk and Technological Culture: Towards a Sociology of Virulence* (Routledge: London, 2002).

[7] Manuel Castells, *The Internet Galaxy* (Oxford University Press: New York, 2002), p. 276.

[8] Rob Shields, *The Virtual* (Routledge: London, 2003), p. 212.

[9] Lee Silver proposes an extropian and horrific solution: the bionic transfer of the sense organs of bats, dogs, spiders, etc. in *Remaking Eden: How Genetic Engineering and Cloning Will Transform the American Family* (Avon Books: New York, 1997). For his part, Gregory Stock sees no likely opposition to such grotesqueries. "To 'protect' ourselves from the future reworking of our biology would require a research blockade of molecular genetics or even a general rollback of technology."—from *Redesigning Humans: Our Inevitable Genetic Future* (Houghton Mifflin: New York, 2002), p. 6.

[10] Boris Groys, "The Insider is Curious, the Outsider is Suspicious," in Geert Lovink, ed., *Uncanny Networks: Dialogues in Virtual Intelligentsia* (MIT Press: Cambridge, 2002), p. 260.

[11] Andrew Feenberg, *Transforming Technology* (Oxford University Press: New York, 2002), p. 190.

[12] D. Urquhart, *Familiar Words* (London 1855), quoted in Marx, Capital I, p. 363.

[13] Adam Smith, *An Inquiry into the Nature and Causes of the Wealth of Nations* [1776] (Modern Library: New York, 1937), pp 734-740.

[14] "The great leap backward," according to Lionel Tiger and Robin Fox, *The Imperial Animal* (Holt, Rinehart and Winston: New York, 1971), p. 126.

[15] Mark C. Taylor, *The Moment of Complexity: Emerging Network Culture* (University of Chicago Press: Chicago, 2001), p. 4.

[16] Katherine Hayles, *How We Became Posthuman* (University of Chicago Press: Chicago, 1998), p. 106.

[17] Hayles, ibid., p. 285.

[18] Hayles, ibid., p. 246. The occasional assertion by such commentators that this reality is at the same time being "highly contested" is the height of irony, as is the mandatory repeated use of the buzzword "body" in virtually every postmodern work of the 1990s.

[19] Very helpful here is Lorenzo Simpson, *Technology, Time, and the Conversations of Modernity* (New York: Routledge, 1995).

[20] Mohammed A. Bamyeh, *The Ends of Globalization* (University of Minnesota Press: Minneapolis, 2000), p. 100.

[21] Scott Lash, *Critique of Information* (Sage: London, 2002), p. 220.

[22] Lash, *ibid.*, p. 1.

[23] I. Bluhdorn, "Ecological Modernisation and Post-Ecologist Politics," in G. Spaargaren, A.P.J. Mohl, and F. Buttel, eds., *Environment and Global Modernity* (Sage: London, 2000).

[24] Steven Best and Douglas Keller, *The Postmodern Adventure: Science, Technology, and Cultural Studies at the Third Millennium* (Guilford Press: New York, 2001), p. 279.

[25] Feenburg, *op.cit.*, pp 4, 189.

[26] Geoffery Hartman, *Scars of the Spirit* (Palgrave/Macmillan: New York, 2002), p. 137.

[27] Bamyeh, *op.cit.*, p. x.

[28] See Theda Skocpol, *Diminished Democracy: From Membership to Management in American Civic Life* (University of Oklahoma Press: Norman, 2003).

[29] Taylor, *op.cit.*, p. 232.

[30] Jean-François Lyotard, *Postmodern Fables* (University of Minnesota: Minneapolis, 1997), p. 31.

[31] Bruce Wilshire, *Fashionable Nihilism: A Critique of Analytical Philosophy* (SUNY Press: Albany, 2002).

[32] Elissa Gootman, "Job Description: Life of the Party" (*New York Times*, May 30, 2003).

[33] Bonnie Rothman Morris, "Lullabies in a Bottle" (*New York Times*, May 13, 2003).

[34] Herbert Marcuse, *Eros and Civilization* (Little, Brown: Boston, 1955), p. 153, for example.

[35] Todd Gitlin, *Media Unlimited* (Metropolitan Books: New York, 2002), p. 163.

13: WHY PRIMITIVISM?

[1] Anselm Giap, *Guy Debord* (Berkeley, 1999), p. 3.

[2] Joseph Wood Krutch, *Human Nature and the Human Condition* (New York, 1959), p. 192.

[3] J. Raloff, "More Waters Test Positive for Drugs," *Science News* 157 (April 1, 2000).

[4] The dramatic upsurge in health-threatening obesity has occasioned many articles, but exact figures are elusive at this time. 27% of adult Americans suffer depression or anxiety disorders. See "Recognizing the Anxious Face of Depression," G.S. Malhi et al, *Journal of Nervous and Mental Diseases* 190, June 2002.

[5] S.K. Goldsmith, T.C. Pellner, A.M. Kleinman, W.E. Bunney, eds., *Reducing Suicide: A National Imperative* (Washington, D.C., 2002)

[6] Claude Kornoouh, "On Interculturalism and Multiculturalism," *TELOS* 110 (Winter 1998), p. 133.

[7] Ulrich Beck, *Ecological Enlightenment: Essays on the Politics of the Risk Society* (Atlantic Highlands, N.J., 1995), p. 37.

[8] Agnes Heller, *Can Modernity Survive?* (Berkeley, 1990), p. 60.

⁹ Michel Houlebecq, *Les Particules Elémentaires* (Paris, 1998). More prosaically, Zygmunt Bauman, *Liquid Modernity* (Cambridge, 2000) and Pierre Bordieu, *Contre-feux: propos pour servirà la résistance contre l'invasion néo-libérale* (Paris, 1998), especially p. 97, characterize modern society along these lines.

¹⁰ Michel Foucault, "What is Enlightenment?" in *The Foucault Reader*, ed. Paul Rabinow (New York, 1984), pp. 47-48.

¹¹ Eric Vogelin, *The Collected Works of Eric Vogelin*, vol. 5, *Modernity Without Restraint* (Columbia, MO, 2000), p.105.

¹² Frederic Jameson, *Postmodernism, or, The Cultural Logic of Late Capitalism* (Durham, NC, 1991), p. ix.

¹³ John Zerzan, "The Catastrophe of Postmodernism," *Future Primitive* (New York, 1994).

¹⁴ Daniel R. White, *Postmodern Ecology* (Albany, 1998), p. 198. Bordieu referred to "the futility of the strident calls of 'postmodern' philosophers for the 'suppression of dualism.' These dualisms deeply rooted in things (structures) and in bodies, do not spring from a simple effect of verbal naming and cannot be abolished by an act of performative magic..."—Pierre Bordieu, *Masculine Domination* (Stanford, 2001), p. 103.

¹⁵ Mike Michael, *Reconnecting Culture, Technology and Nature* (London, 2000), p. 8. The title itself is testimony to the surrender to domination.

¹⁶ Iain Chambers, *Culture After Humanism* (London, 2002), pp. 122, 41.

¹⁷ Recent titles in various fields indicate a shift. For example, *Calvin O. Schrag and the Task of Philosophy After Postmodernity*, eds. Martin Beck Matustic and William L. McBride (Evanston, IL, 2002) and *Family Therapy beyond Postmodernism* by Carmel Flaskas (New York, 2002). *After Poststructuralism: Writing the Intellectual History of Theory*, eds. Tilottama Rajan and Michael J. Driscoll (Toronto, 2002) is haunted by themes of origins and the primitive.

¹⁸ Jean-François Lyotard, "Domus and the Megalopolis" [which could very legitimately have been called, in anti-postmodernist fashion, "From Domus to the Megalopolis"] in *The Inhuman: Reflections of Time* (Stanford, 1991), p. 200.

¹⁹ Lyotard, *The Inhuman*, p. 200, and *Postmodern Fables* (Minneapolis, 1997), p. 23.

²⁰ Michael Hardt and Antonio Negri, *Empire* (Cambridge, MA, 2000).

²¹ Hardt and Negri, p. 218.

²² Claude Kornoouh, "Heidegger on History and Politics as Events," *TELOS* 120 (Summer 2001), p. 126.

²³ Oswald Spengler, *Man and Technics* (Munich, 1931), p. 94.

²⁴ Spengler, *Man and Technics*, p. 69

²⁵ Spengler, *Früzeit der Weltgeschichte*, #20, p. 9. Quoted in John Farrenkopf, *Prophet of Decline* (Baton Rouge, 2001), p. 224.

²⁶ Spengler, *Man and Technics*, p. 103.

²⁷ Theodor W. Adorno, *Prisms* (London, 1967), p. 72.

²⁸ Max Horkheimer and Theodor Adorno, *Dialectic of Enlightenment* (New York, 1947).

²⁹ Horkheimer and Adorno, *Dialectic of Enlightenment*, p. 55.

³⁰ Albrecht Wellmer, *Endgames: the Irreconcilable Nature of Modernity* (Cambridge, MA, 1998), p. 255.

³¹ Herbert Marcuse, *Eros and Civilization* (Boston, 1955).

³² Sigmund Freud, *Civilization and its Discontents* (New York, 1961).

³³ Emile Durkheim, *The Division of Labor in Society* (New York, 1933), p. 249.

³⁴ Sigmund Freud, "The Future of an Illusion," *The Complete Works of Sigmund Freud*, vol.

21 (London, 1957), p. 15.

[35] Important texts include Eleanor Leacock and Richard B. Lee, *Politics and History in Band Societies* (New York, 1982); Richard B. Lee and Richard Daly, *The Cambridge Encyclopedia of Hunters and Gatherers* (Cambridge, 1999); Marshall Sahlins, *Stone Age Economics* (Chicago, 1972); Colin Turnbull, *The Forest People* (New York, 1968); Adrienne Zihlman, et.al., *The Evolving Female* (Princeton, 1997).

[36] M.J. Morwood, et. al., "Fission-track ages of stone tools and fossils on the east Indonesian island of Flores," *Nature* (12 March 1998), for example.

[37] This tendency within an increasingly anarchist-oriented anti-globalization movement is in the ascendant in the U.S. Among a growing number of periodicals are *Anarchy, Disorderly Conduct, The Final Days, Green Anarchy, Green Journal*, and *Species Traitor*. Texts include Chellis Glendinning, *My Name is Chellis and I'm in Recovery from Western Civilization* (Boston, 1994); Derrick Jensen, *Culture of Make Believe* (New York, 2002); Daniel Quinn, *Ishmael* (New York, 1995); John Zerzan, *Running On Emptiness: the Pathology of Civilization* (Los Angeles, 2002).

[38] Joel Whitebook, "The Problem of Nature in Habermas, " *TELOS* 40 (Summer, 1979), p. 69.

[39] Cornelius Castoriadis, *Crossroads in the Labyrinth* (Cambridge, MA, 1984), p. 257. Also, Keekok Lee, "To De-Industrialize—Is It So Irrational?" in *The Politics of Nature*, eds.. Andrew Dobson and Paul Lucardie (London, 1993).

[40] George Grant, *Technology and Empire* (Toronto, 1969), p. 142. Of course, the situation grows more and more grave, with sudden, dire changes very possible. M. Sheffer, et. al., "Catastrophic Shifts in Ecosystems," *Nature* (11 October 2001); M. Manion and W.M. Evan on the growing likelihood of disasters, "Technological Catastrophes: their causes and preventions," *Technology in Society* 24 (2002), pp. 207-224.

[41] Claude Kornoouh, "Technique et Destin," *Krisis* 34 (Fall, 2000).

14: SECOND-BEST LIFE: REAL VIRTUALITY

[1] Slavoj Zizek, *The Plague of Fantasies* (New York: Verso, 1997), p. 44.

[2] Allucquere Rosanne Stone, "Will the Real Body Please Stand Up?" in Michael Benedikt, ed., *Cyberspace: First Steps* (Cambridge, MA: MIT Press, 1991).

[3] Joseph Hart, "Grief Goes Online" in *Utne*, April 2007.

[4] Wade Roush, "Second Earth" in *Technology Review*, July/August 2007, p. 48.

[5] Widely circulated books include: Howard Rheingold, *Virtual Reality* (New York: Summit Books, 1991), Michael Heim, *The Metaphysics of VR* (New York: Oxford University Press, 1993), Rudy Rucker, R.U. Sirius, Queen Mu, *Mondo 2000: A User's Guide* (New York: Harper-Collins, 1992), Nadia Magnemat Thalmann and Daniel Thalmann, *Virtual Worlds and Multimedia* (New York: Wiley, 1993), Benjamin Woolley, *Virtual Worlds* (Cambridge, MA: Blackwell, 1992). An excellent corrective is Robert Markley, ed., *Virtual Realities and Their Discontents* (Baltimore: Johns Hopkins University Press, 1996).

[6] For his idiosyncratic twist on this, see Jean Baudrillard, *Forget Foucault* (New York: Semiotext, 1987).

[7] Philip Zai, *Get Real: A Philosophical Adventure in Virtual Reality* (Lanham, MD: Rowman & Littlefield, 1998), p. 171.

[8] Martin Heidegger, "Nietzsche's Word: 'God is Dead'" in his *Off the Beaten Track*, translated and edited by Julian Young and Kenneth Haynes (Cambridge: Cambridge University

Press, 2002), p. 191.

⁹ Hans-Georg Gadamer, *Philosophical Hermeneutics*, translated and edited by David E. Linge (Berkeley: University of California Press, 1976), p. 71.

¹⁰ Bill Brown, ed., *Things* (Chicago: University of Chicago Press, 2004); Stephen Melville, ed., *The Lure of Things* (Williamstown, MA: Sterling and Francine Clark Art Institute, 2005).

¹¹ Brown, *op. cit.*, p. 12.

¹² Heinz R. Pagels, *The Dreams of Reason: The Computer and the Rise of the Sciences of Complexity* (New York: Simon and Schuster, 1988).

¹³ Oswald Spengler, *Man and Technics*, translated by Charles Francis Atkinson (Westport CT: Greenwood Press, 1976), p. 94.

¹⁴ David Gelerntner, *Mirror Worlds* (New York: Oxford University Press, 1991), p. 34.

¹⁵ Ludwig Wittgenstein, *Culture and Value*, translated by P. Winch (Oxford: Blackwell, 1986), p. 56.

15: BREAKING POINT?

¹ Frank Furedi, *Culture of Fear* (London: Cassell, 1997).

² Miller McPherson, Lynn Smith-Levin, Matthew E. Brashears, "Social Isolation in America"in *American Sociological Review*, 2006, pp. 353-395.

³ Dr. Kunio Kitamura, quoted in *The Japan Times* (Reuters, June 23, 2006). Also, Edwin Karmiol, "Suicide Rate Takes a Worrisome Jump" (*Asia Times*, August 4, 1999).

⁴ Theodor Adorno, *Negative Dialectics* (New York: Continuum, 1997), p. 267.

⁵ Valovic, *Digital Mythologies*, 2000, p. 178.

⁶ Daniel Downes, *Interactive Realism: The Poetics of Cyberspace* (Montreal: McGill-Queens University Press, 2005), p. 144.

⁷ Melissa Holbrook Pierson, *The Place You Love is Gone* (New York: W.W. Norton, 2006), p. 85).

⁸ Min Lin, *Certainty as a Social Metaphor: The Social and Historical Production of Certainty in China and the West* (Westport, CT: Greenwood Press, 2001), p. 133.

⁹ Jean-Francois Lyotard, *The Inhuman: Reflections on Time* (Stanford: Stanford University Press, 1991), p. 63.

¹⁰ Jacques Ellul, *The Technological Society* (New York: Alfred A. Knopf, 1964), p. 5.

¹¹ Lyotard, *op. cit.*, p. 12.

¹² Koert van Mensvoort, *Exploring Next Nature* (http:www.koert.com)

¹³ Paul Berkett, *Marxism and Ecological Economics: Toward a Red and Green Political Economy* (Boston: Brill, 2006), p. 328.

¹⁴ Sigmund Freud, *Civilization and Its Discontents*, p. 99.

¹⁵ Adorno, *op. cit.*, p. 323.

16: FINDING OUR WAY BACK HOME

¹ Czeslaw Milosz, *The Land of Ulro* (New York: Farrar, Strauss, Giroux, 1985), p. 229.

² Earle H. Waugh and K. Dad Prithipaul, eds., Native Religious Traditions, "Mohawk Seminar" (Waterloo, Ontario: Canadian Corporation for Studies in Religion, 1977), p. 37.